Fireplaces, Chimneys & Stoves

Fireplaces, Chimneys & Stoves

Michael Waumsley

THE CROWOOD PRESS

First published in 2005 by
The Crowood Press Ltd
Ramsbury, Marlborough
Wiltshire SN8 2HR

www.crowood.com

British Library Cataloguing-in-Publication Data
A catalogue record for this book is available from the British Library.

ISBN 1 86126 746 0

Disclaimer
The author and the publisher do not accept responsibility in any manner
whatsoever for any error or omission, nor any loss, damage, injury, adverse
outcome, or liability of any kind incurred as a result of the use of any of the
information contained in this book, or reliance upon it. Readers are advised to
seek specific professional advice relating to their particular house, project and
circumstances before embarking on any installation or building work.

Typeset by Jean Cussons Typesetting, Diss, Norfolk

Printed and bound in Singapore by Craft Print International

Contents

Preface

This book is not intended as a textbook on the construction of chimneys or the installation of a fireplace or stove – the skills needed for these tasks cannot be learnt in this way and can only be perfected as a result of training and practice. The aim of the book is to ensure that when the reader decides to have a fireplace, stove or chimney fitted or built, they have the information they need to be able to make informed choices.

With such a large number of choices to be made it can be quite daunting for a new user, and the help and advice offered in the book can make all the difference between being satisfied or disgruntled with the final result. There is a brief history of the fireplace to enable consideration of the period of the property to be taken into account; while this should not be the main reason behind a choice, it can be a factor in it. Do you go for a closed or open appliance; inset or free-standing installation; traditional or contemporary style; electric, gas, oil or solid fuel? These are all important factors and must be settled before any actual purchases are made.

These and other factors could be governed either by what type and size of chimney is present and what condition it is in, or by what type of flue or chimney you intend to construct.

When the final choice has been made, details on how to gain the required permissions, and the reasons for testing and surveying are included. Information is then provided to enable the reader to tell if the installation is being carried out correctly, in order to help ensure both safety and a satisfactory result. This follows the old maxim that to be forewarned is to be forearmed.

ACKNOWLEDGEMENTS

I would like to thank all my friends and colleagues in the solid fuel and chimney industries for their help and support in this project, especially those manufacturers who have kindly allowed me to use their illustrations and photographs in this book.

Of course, this project has taken a considerable time away from my family and so I would like to thank them for their patience and support.

I would like to dedicate this book to two young ladies who, despite being still below school age, have suffered the kind of troubles and difficulties in their young lives that I can only thank God I have never had to suffer, but who have provided me with inspiration and amusement whenever I felt things were getting difficult; may their laughter and smiles prove an inspiration for many years to come. Keep it up, Olivia and Millie.

Introduction

Fire has been a major part of the human survival kit almost since the beginning of the human race and as such has provided warmth and light as well as cooking and protection. Even today the modern central heating system in your home is likely to rely on combustion as the main source of its energy at some point in the generation of heat.

In the modern property the generation of heat for a heating system is normally clinical, efficient and hidden. This modern way has resulted in the loss of the spiritual element of fire and so a primeval source of well being is missing from most of our lives. In more recent times, the yearning for the element of fire in our lives has resulted in an upsurge in the popularity of a fire as a focal point of our main living space.

One of the issues brought about by the period of not having fire as the focal point in the living room has been the loss of the skills needed to construct a safe and correctly operating fire, stove and chimney installation. The prime aim of this book, therefore, is to ensure that anyone considering the fitting of a fire in their current home, or the construction of a chimney in a new home, can find all the information they need, both technical and aesthetic, in the one place.

As a beginning of the information trail I intend to provide a short insight into the history of fire and chimneys within our homes, from the simple hole in the roof or wall to the modern system-built chimneys and brick structures that are based on well-developed scientific principles. I will also provide an insight into the styles of fires that were in common use at differing periods of housing construction to enable those who wish to match the fireplace they install, or

the chimney they construct, to the period or style of home they have. However, it is important not to get too carried away with the need to match the focal point of the room with the style of the house – it is more important to choose a fire or stove that will give you years of pleasure, both safely and efficiently, within the budget you have available.

The fires used in our homes throughout the ages have varied from a large structure in which a child could stand next to the fire in the opening in order to turn the roasting spit, through to smaller open fires, high-efficiency stoves and modern central heating boilers. All fires were originally fuelled by wood from the forest, but we have passed through periods of coal burning, the use of smokeless fuels and the development of near perfect combustion of gas and liquid fuels. The most popular types of focal-point fires and stoves these days have seen a return to the burning of wood. Wood is now recognized as the most environmentally friendly combustion-appliance fuel available today. A tree will absorb more carbon monoxide in its growth than it will create when burnt and with modern combustion techniques it is possible to create near-complete combustion. A complete tree can be burnt to leave little more than a few pounds of ash and of course if the forests are managed correctly we can have a whole new fuel supply within fifteen to twenty years. It is hoped that the information given on the different types of appliances will lead the reader to the perfect choice of appliance for their needs and desires by providing comparative data, although at the end of all this careful consideration the end decision will almost certainly be based on a feeling from the heart.

The book goes on to discuss the safety aspects of the installation and the need for the correct combination of fire, fuel and flue as well as all the other safety-related issues. It is a sad fact that the UK still has an unacceptably high incidence of carbon monoxide poisonings from combustion appliances and it has been my much regretted task to have to investigate a number of these incidents in the past. Not only can these incidents be life-changing, but they are also life-threatening, and so it is important that whatever information you take away with you from reading this book the safety section is at the forefront whenever a decision is made.

In the UK the regulations relating to the installation of combustion appliances, chimneys and flues have all recently been completely overhauled, bringing in safer practices and a more detailed and demanding set of requirements. This is intended as a major step in improving the safety of householders and reflects the growing trend towards the inclusion of chimneys in new houses as well as the opening up of disused flues in older ones. The technical information contained in this book has a large section intended to cover the requirements of building control and planning, as well as the British and European Standards that have to be adhered to, or that provide additional information for our consideration. To help with the understanding of these technical documents and the information therein, a glossary of terms is included at the back of the book.

This book aims to cover all the information that a customer needs when engaging a professional to carry out any work on a fireplace or chimney, as well as if the reader decides to go it alone. There is information to cover the aspects of the design, planning and construction of a chimney and flue as well as the choice of heating appliance, fuel and decorative surround. What this book will not attempt to do is teach the reader the skills needed to lay the bricks or shape the stone and so on; these can only be developed as a result of training and practice with the actual tools and materials and cannot be obtained purely by the reading of a book.

USING THIS PUBLICATION

This book has been written to provide a path through the process of deciding what it is you require from the fire you are planning to fit; it gives information and advice for the types of appliances available to guide you through the choice of the appliance and then advice on the flue required and the installation process. Even if you do not intend to install the appliance yourself it is important to follow a process and have background knowledge to ensure the resulting installation is carried out correctly, conforms to the requirements of the law and is safe for you to use.

If you continue through this process and still have a problem that requires some analysis, then towards the rear of the book there is advice on basic fault diagnosis to help you to identify the fault and recommend a solution.

I hope that the information in the following chapters will provide the details you need in order to be able to decide on the most suitable type of fire and installation, but it is important to remember that there are always issues that the professional may pick up that an inexperienced person could not, and so although using the services of a competent person may seem expensive, it can save a lot of time, effort and risk in the long run.

CHAPTER 1

The History of Fire in Our Homes

This chapter is not intended as a history lesson about our ancestors' needs and desires, but more as a warning when choosing a fire for your own home. It is often the case that we see a fireplace or stove that becomes our heart's desire while we are visiting an ancestral home or one of the many National Trust properties. What we see before us is a beautiful period fireplace or a large rustic fire with an impressive oak beam. What we fail to see due to the high standard of care and restoration carried out in these properties are the stains on the chimney breast or the charring of the beam. Other issues we have to consider are whether the windows would have had glass in them or just a drape to keep out the draught, or whether the chairs would have had high backs and perhaps sides to protect the lord of the manor from the draughts created when the servant lit the fire.

It is essential that we also consider the conditions in which historical fireplaces were conceived and used, and whether we would be happy, or even safe, living with these conditions in our modern, comfortable and well-sealed homes.

THE EARLY USE OF FIRE IN OUR HOMES

It is unlikely any reader has a cave as their dream home and so I do not need to go back this far; we will therefore take the central hearth as our starting point. It is often understood that these were accompanied by a hole in the roof to allow the smoke to exit. In reality, the smoke would fill the roof and higher parts of the hall and exit through the windows (which were

only holes in the walls with drapes), as most of the smoke would be unable to find the hole provided in the roof.

Fig 1   Illustration of a lantern over a hole in roof.

In the more elaborate properties with central hearths the hole in the roof would be larger, but it would be protected by a lantern. The lantern served two main functions – one was to stop the rain and snow entering the hall and the second was to create a scary noise in bad weather to ward off evil spirits. In Portugal and some other Mediterranean countries lanterns are still used on chimneys for these reasons. Some say this is the first chimney, but it is the author's opinion that they are the first chimney pots. A chimney contains a flue which connects the fire to the outside atmosphere, and in this case no structure or flue existed. It was difficult for this structure to be safely provided until the fire was moved to the side of the room up against a wall, in the eleventh and twelfth centuries.

In these early days this would not be a chimney as we know it, but a simple vent in the wall just above and to the back of the fire. If you were to try this today you would quickly realize that at best it would remove only a very small amount of the smoke created and, at worst, when the wind blew in the vent, it would ensure that all the smoke came back into the hall.

As the fireplace and the chimney became more developed, the position of the vent moved higher until eventually the chimney containing the flue reached the roof. One of the first chimneys we would recognize as such was constructed on the Jew's Palace in Newark, Nottinghamshire. In this case, the chimney became a structure that projected above roof level and was taken to a position where it would be less adversely affected by the wind. Even at this time, although the fireplaces had begun to operate more successfully than in the past, they still did not get rid of all of the smoke from the room because they were so large that not all the smoke could be retained within them.

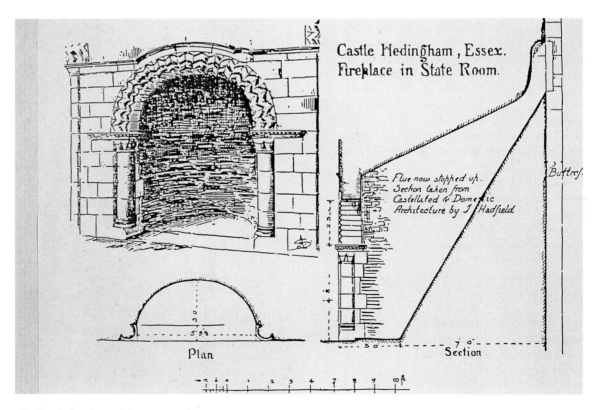

Fig 2   A fireplace with a vent at the rear.

*Fig 3   Early
flues on buildings
in Southall,
Nottinghamshire.*

At this time, many older properties had the central hearth removed and the fires placed up against an outside wall so that a chimney and flue could be constructed. This also allowed the insertion of intermediate floors. There are many examples in almost all parts of Britain of chimneys being added onto older properties. The chimney became a common sight on new properties, and the fire developed into something seen not only in the main hall of a property, but in the homes of the better-off owner they also became common in the bedrooms.

These fireplaces never worked in the way we would expect today and allowed a large quantity of smoke and fumes to enter the room. This was not as dangerous as it is today because the levels of ventilation were considerably greater – there was no glass in the windows and the wind would enter the hall almost unhindered. The lord at this time would have to sit in front of the fire in a chair with a high back and sides to protect him from the draughts caused when the fire was lit and he would be on a raised platform or have a footstool to raise his feet out of the cold air moving along the floor.

*RIGHT: Fig 4   A diagram of chimney at the
Jew's Palace, circa 1350.*

Fig 5   A brick flue added to a stone property.

Fig 6   A brick flue on the top of a stone wall.

The fireplaces in bedrooms gradually became smaller. These were found to operate more effectively than a large fire, and so from the sixteenth century onwards even the main fires in large rooms began to be constructed in such a way that the fireplace and overmantel was the dominant feature and the actual opening for the fire became smaller. Larger fireplaces were less efficient, caused a greater air movement and so more draughts, while still allowing a high proportion of the smoke back into the room. As time passed, metal inserts were added to reduce the fire size but provide a good view of the fire; this would also get hot and radiate additional heat into the room.

By 1796 it was common for the Rumford fireplace to be fitted. These can still be purchased today from Buckley Rumford Fireplaces, and this type of unit was the first to include the angling of the side walls of the fireplace. This was done because Count Rumford understood that the heat generated by an open fire is radiant heat and the best way to obtain a high heat output from an open fire is to ensure that as much of the radiant heat as possible escapes the fireplace.

The Rumford fireplace was designed to reflect the heat into the room as opposed to the cast-iron fire insert absorbing the heat and then radiating it back out. This process would primarily serve to warm the air passing over the cast iron, which would ultimately be drawn up the chimney and lost to atmosphere.

To assist in this reflection of heat Count Rumford advocated the painting of the side 'reflecting walls' white to help the reflection. Modern thinking is that Plank's Law indicates that what really happens is that these side walls, painted or not, do absorb the heat

Fig 7   A large fireplace and mantel, circa *seventeenth century.*

Fig 8   A register hob grate.

Fig 9   A horseshoe register grate.

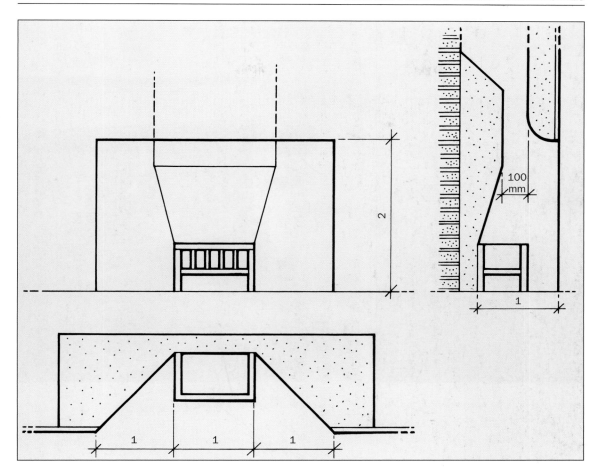

Fig 10   Rumford fireplace ratios. These are the ratios that Count Rumford calculated gave the best evacuation of flue gases and heat into the room.

and then re-radiate it into the room and that there is little difference in the heat from a painted wall or a sooty black one.

The additional change to the fireplace instigated by the Rumford design was to smooth off the route the gases take into the chimney and to create a venturi (known as the throat in the case of an open fire), which speeded up the gas flow at the critical point and then injected the smoke into the gather or smoke chamber above. This restriction also reduced the amount of heat lost up the chimney and greatly reduced the incidence of smoking back. The basic principles used in the Rumford fireplace are still current today in the British Standard Fireplace components, with some modifications to bring the throat forward and the fire basket further back.

If a large fireplace opening is chosen, then a study of the Rumford design would enable more heat output, cleaner burn and fewer emissions as well as reducing the air entering the room to ventilate the fireplace.

THE MORE RECENT USE OF FIRE IN OUR HOMES

By this stage, people had started to use the fire for

Fig 11   A cooking range from the nineteenth century, still available in reproduction form from the Yorkshire Range Company.

Fig 12   A tiled fireplace from the 1930s–1950s.

cooking in a different way due to the development of the range cooker. This became a common sight in many homes, and can still be seen in some older homes today. If this is the type of fire you require today they can be obtained either refurbished or as a new unit. The Yorkshire Range Company still produces a full cooking range with open fire today.

The now familiar standard fireplace with the shaped fireback and ceramic tiled fire surround became the norm in the 1930s and remained so well into the 1950s and 1960s. However, it was very much altered in the 1970s when we began experimenting again with different styles and shapes.

In the late 1950s and 1960s the advent of central heating and the use of gas and electric provided a convenient alternative to solid fuel and the use of a cheaper and simpler flue took over from the chimney. From 1967 the need for a chimney to be lined increased the cost of construction and so the building of new houses with chimneys began to decline, until most were constructed without a chimney. However, this was not the end of the story, as during the 1990s a fire in the living room became a fashion statement and many of the old chimneys were brought back into use. The next stage was for houses to be built with chimneys capable of accepting a solid fuel fire, and in the detached house market this is rapidly becoming the norm.

This process is continuing today, with many of the properties built in the 1960s–1980s having new chimneys added as the owner wishes to have a fire-

Fig 13   A hole-in-the-wall fire from the 1970s.

place installed. To add still further to the demand for a solid fuel fireplace it is now possible to obtain stoves that are 80 per cent efficient. This type of stove is seen as the most environmentally friendly source of heating using a combustion process.

CHAPTER 2

Safety

Every type of fuel to be burnt in the home will have specific risks and if you are to 'reduce these risks to a reasonable level', as is required by the Building Regulations, then consulting the regulations, standards and installation instructions for any product is essential.

The safety of the occupant of any house has to be the prime concern when any work is carried out to that property, even more so when dealing with something as dangerous as fire and smoke. The dangers of fire will be well understood by almost any reader, but can still be underestimated due to the long history we have of containing and controlling fire in our homes.

One of the greatest dangers to occupiers of homes containing a combustion appliance is carbon monoxide (CO) poisoning. This gas is highly toxic and has become known as the silent killer; information on this gas is therefore contained in this chapter.

FIRE PRECAUTIONS

The danger of fire in a house is well understood by those who have to deal with it and those who have to design to avoid it. In this case we are designing to contain it, which in many ways produces a greater risk. The risk of fire in our home will vary dramatically depending on issues such as the type and controllability of the fuel supplied to the fire, the efficiency of the appliance, the appliance type, the level of use and the quality of maintenance.

All the regulations in place are there to provide advice and information on how to contain the fire safely within the fireproof chamber we know as the fireplace and chimney. These regulations and standards may at first sight seem restrictive and just another example of red tape, but it is imperative that they are followed and that work is carried out by a competent person – if not, the consequences could prove tragic or even fatal.

There are a number of dangers that are often ignored or not considered, including chimney fires from a wood-burning stove, and the heat transference into the floor under a gas fire and through the back wall of the fireplace where an open fire is in use. If a chimney fire occurs in a chimney that has developed a coating of tar on the inside, this could burn for a few hours and reach temperatures of over 1,000°C. This is sufficient to cause a steel pipe to glow red and brick and cement to crack and degrade. Even ignoring this extreme, if we consider only the normal operating temperatures in the flue we will underestimate the dangers and put ourselves at risk. Building Regulations Section J and the relevant Approved Document, when read with British Standards, provides advice on the location of flue pipes in relation to combustible material, the size and type of hearth that should be present as well as what can be placed on the wall behind a fireplace and how thick that wall should be.

CARBON MONOXIDE

Carbon monoxide (CO) is a colourless, odourless and tasteless gas that is the product of incomplete combustion of carbon-based fuels such as smokeless fuels, coal, gas, oil and even wood. It is harmful if breathed in and may result in death if present in large enough quantities or if a person is exposed to it for long enough.

The Effects of Carbon Monoxide

When carbon monoxide is breathed in it reacts with haemoglobin (part of the red blood cell) to form a compound called carboxyhaemoglobin. The bond between the CO and haemoglobin is far stronger than between oxygen and haemoglobin and so does not release easily and remains in the blood for a long time. Since haemoglobin is the part of the blood that normally carries oxygen around the body, if a lot of it is contaminated then the quantity of oxygen in the system will be reduced. If the body continues to be exposed then eventually there will not be enough oxygen being transported around the body and death will occur.

The connection of CO and haemoglobin is not permanent, so sudden high doses of CO are normally reversible. If the victim is removed to a CO-free atmosphere quickly enough, they are likely to make a complete recovery. However, the bond that forms carboxyhaemoglobin lasts for much longer than that between oxygen and haemoglobin, and so long-term exposure to low levels of CO will cause CO poisoning. Each time CO is breathed in, a little more haemoglobin becomes contaminated and the harmful effects accumulate in the body. Long-term exposure to low levels of CO will therefore lead to high levels of carboxyhaemoglobin in the blood. These levels can attack the proteins in the body and cause muscle degradation and damage to vital organs, including the brain; this damage cannot be reversed even when there is no carboxyhaemoglobin remaining in the blood.

The effects of carboxyhaemoglobin in the blood are influenced by the following factors:

* the concentration of CO in the inhaled air;
* the duration of exposure to the contaminated air;
* the degree of activity of the person in the contaminated atmosphere (the harder a person works, the greater his or her breathing rate and the higher the absorption);
* individual susceptibility (people with heart or lung disorders, young children, the aged and pregnant women may all be at increased risk).

Symptoms and Warning Signs

The physical symptoms that will be experienced by a person exposed to CO vary with the level of

Contaminated Haemoglobin Percentage	Possible Symptoms
0–10	None
10–20	Tightness across the forehead
20–30	Headaches
30–40	Severe headaches, weakness, dizziness, nausea and vomiting
40–50	Collapse, increased pulse and respiratory problems
50–60	Coma, intermittent convulsions
60–70	Depressed heart action, death possible
70–80	Weak pulse, slowed respiration, death likely
>80	Death in minutes

contamination of the blood. The table on page 19 is a summary of these symptoms in table form. The level of contamination is expressed as the percentage of haemoglobin in the blood that has been converted into carboxyhaemoglobin.

The early symptoms of CO poisoning are similar to those of common flu and in the latter stages of low-level poisoning, muscle wastage, loss of memory and concentration, confusion and weakness all have many other possible causes and are not always easy to diagnose. It is also often stated that the skin will go pink in a CO poisoning victim, but this is only present in around 50 per cent of cases.

It is often commented that the complainant feels better when they are away from home but relapses on return. Another indicator is if more than one person in a property is suffering the same symptoms at the same time. *If you are in any doubt you should seek medical advice.*

Sources of Carbon Monoxide

The most significant sources of CO include motor-vehicle exhaust fumes and the burning of carbon- and hydrocarbon-based fuels such as coal, smokeless fuel, oil and gas (although cigarette smoking is another source). The use of portable gas and paraffin heaters in poorly ventilated, confined spaces can result in high levels of CO in the atmosphere. All fires, room heaters, boilers and cookers fuelled by natural gas, liquid petroleum gas (LPG) or oil require an adequate supply of air (*see* next section) to ensure complete combustion and minimal formation of CO.

In the case of solid fuels, that is, coal, wood and smokeless fuels, it is not possible to create perfect combustion as the fuel will not mix with oxygen in the same way as will gaseous and pressurized liquid fuels. For this reason, the gases from these fuels will always contain significant (and dangerous) levels of CO, which can rapidly cause levels of several thousand parts per million of CO in a room if a flue or air vent is blocked. In the case of these fuels it is essential that a correctly operating flue transports all the smoke and fumes out to atmosphere. This normally requires a greater level of ventilation than for other fuels and chimneys and flues must be regularly swept and meticulously maintained.

AIR FOR COMBUSTION AND VENTILATION

All combustion appliances require an adequate supply of air. The supply of air to a solid fuel, or other open flue appliance has two functions:

- to provide oxygen for combustion; a flame in a sealed container will extinguish as soon as all the available oxygen has been used up;
- to provide the additional air required to enable the products of combustion to be ventilated through the chimney flue to the atmosphere.

Air for Combustion

For the combustion of any fuel to take place oxygen is required. With improved standards of construction and draught-stripping, the amount of advantageous air (air introduced through gaps around doors, windows, floorboards and so on) has dropped significantly. For this reason, the Building Regulations *Approved Document J* (*ADJ*) now gives guidance on the amount of fixed air vents needed.

Some solid fuel open fires (underfloor draught types) provide air for combustion directly to the under-grate area of the fire from outside. This is air for combustion and amounts to only 10 per cent of the total air requirements. The other 90 per cent is required for ventilation.

Air for Ventilation

Air is required for ventilation in addition to the air needed for combustion. This is the air that is used to replace that drawn into the chimney to transport the products of combustion to the atmosphere. If there is insufficient ventilation air, the products of combustion will travel more slowly up the flue and so deposit soot and condensation. If the lack of ventilation (air starvation) becomes acute, 'fuming' or 'smoking back' may occur, which could lead to carbon monoxide poisoning.

Open fires and stoves that can be operated with their doors open will require far more air for ventilation than for combustion. In the case of the large period fireplaces or inglenooks the volume of air in a room may be replaced many times in just one hour.

LARGE OPEN FIRES

The term 'large open fire' is in this case meant to define a fireplace larger than those specified in the British Standards and which does not have a throat formed using British Standard components.

The opening size of an open fire will dictate how much air will be needed to ensure the complete evacuation of the dangerous products of combustion. The larger the fireplace opening, the more air it will need to make it operate correctly.

When smoke is produced it wants to rise up the chimney. Not all of the smoke will stay in the fireplace unless the chimney also draws air in from the room to drag the smoke back into the fireplace. The air must enter the fireplace with a given average speed over the whole of the fireplace to ensure none of the smoke escapes into the room. The greater the fireplace size, the more air will have to be drawn into the chimney to be able to achieve this speed.

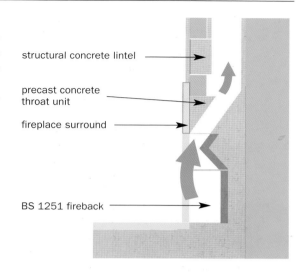

structural concrete lintel

precast concrete throat unit

fireplace surround

BS 1251 fireback

Fig 14   The flow of air into a fireplace and flue must be from the room in order to drag the smoke back into the flue.

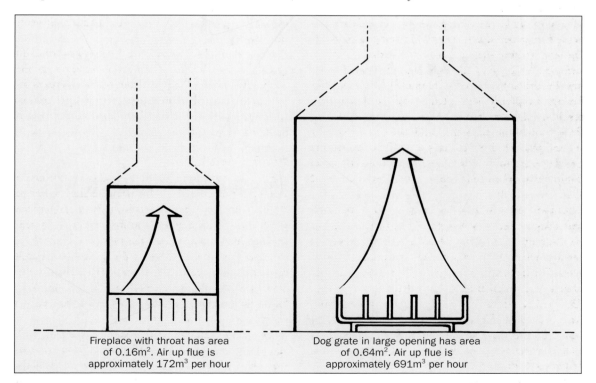

Fireplace with throat has area of 0.16m². Air up flue is approximately 172m³ per hour

Dog grate in large opening has area of 0.64m². Air up flue is approximately 691m³ per hour

Fig 15   Air is drawn from the room – the bigger the fireplace, the more air is taken up the chimney.

Examples of Air Requirements for Different Fireplace Opening Sizes

- Fireplace height = 0.55m (22in).
- Fireplace width = 0.5m (20in).
- Fireplace area = 0.55 × 0.5 = 0.275m^2 (3sq ft).

With an average speed of 0.3m/s the volume of air entering the fireplace will be 0.275 × 0.3 = 0.0825m^3/s, which is the same as 297m^3/hour (10,500cu ft per hour).

In the case of a larger fire, let us consider a modest 1.0 × 1.0m (39 × 39in) fireplace; the calculation will be as given below:

- Fireplace height = 1.0m (39in).
- Fireplace width = 1.0m (39in).
- Fireplace area = 1.0 × 1.0 = 1.0m^2 (10.5sq ft).

With an average speed of 0.3m/s the volume of air entering the fireplace will be 1.0 × 0.3 = 0.3m^3/s, which is the same as 1,080m^3/hour (37,000cu ft per hour).

It is difficult to imagine a quantity of air equivalent to the volume of a room 19 × 19 × 3m high (62 × 62 × 9ft high) travelling through your living room every hour, but that is what will be required if this fire is to work safely.

To provide this air it is necessary for there to be air inlets into the room, preferably in a convenient location to avoid the air movement causing discomfort. This part of the system must therefore be given considerable thought at an early stage.

A fireplace of 1 × 1m (39 × 39in) in a centrally heated living room is likely to have zero efficiency and may, if air vents are badly positioned, have negative efficiency. In this case, you may feel warmer sitting in front of the fire while it is lit, but the temperature in the corners of the room may drop as the central heating struggles to keep all this air movement warm.

SIMPLE OPEN FIRES

This is a term used to describe an open fire built using the British Standard components. It includes a good throat and is normally no more than 550 × 450mm (21½ × 17½in).

This type of fire is based on the Count Rumford design. A good throat formation and gathering are essential for effective clearance of the products of combustion. The restriction at the throat increases the initial velocity of these gases into the flue and prevents smoke spillage into the room, whilst, at the same time, reducing the amount of airflow through the room and the unnecessary heat loss up the flue.

The throat should be 110mm (4in) from the front to the back and about 300mm (12in) wide. Many existing inset fires will be constructed with a British Standard throat-forming lintel in conjunction with the traditional corbelled brickwork gather.

PROVIDING VENTILATION

The required ventilation can be provided in many ways: direct from outside; through an outside wall; through a ventilated floor; or via another room that is ventilated. Any vent installed must be permanently open and not capable of being closed.

When selecting an air vent it is important to ensure it has sufficient open area. Approved Document J of the Building Regulations gives the air requirements for different types of appliances and this information is included in Figure 16, but the actual requirement for any particular appliance must be checked in the installation instructions, as some may require more, or allow the fitting of less, ventilation.

Fig 16 The air supply to an open-flued appliances in fireplaces, including solid fuel, gas and oil.

Fuel Type	Appliance type	Classification	Minimum vent size (cm²)
Gas	Decorative fuel effect gas fire	For fireplace openings with throats	100
		For fireplaces with no throat or under a canopy calculate vent size as for a solid fuel open fire	50% of the flue area serving the appliance
		For fires with less than 7kW input and a flue gas clearance rate of less than 70m³ per hour	0
	Inset Live Fuel Effect Fires and radiant gas fires	Up to 7kW input	0
		Above 7kW input	500mm² per kW of input above 7kW
Oil	all open flue appliances	Up to 5kW output	0
		Above 5kW output	550mm² per kW of output above 5kW
Solid fuel	Open Fires	A fire fitted to BS 8303 with BS 1251 components	50% of the throat area
		A fire in a recess without a throat or one fitted under a canopy	50% of the flue area serving the appliance
	Closed Appliance	One that can be used with the fire doors open	50% of appliance offtake area (this acts as a throat)
		One designed only to be used with the fire doors shut up to 5kW output	0
		One designed only to be used with the fire doors shut above 5kW output	550mm² per kW of output above 5kW

Terracotta airbricks do not give as large a free area as may be expected – for example, a traditional 215 × 140mm (9 × 5½in) airbrick will only give around 4,570mm² (7.5sq in). An equivalently sized plastic air vent will provide up to 12,000mm² (20sq in). It is important to note that the fitting of a fly screen (a mesh with small holes to prevent insects from entering) will greatly reduce the free area of a vent and in some cases may be against the Regulations.

SITING OF VENTS

From the point of view of air supply it does not matter where the air vent is sited, but for an occupant in the room it can be very important. It is recommend that when positioning an air vent its position should minimize cold draughts, not be unsightly and be in a position where it will not accidentally become blocked.

For a solid fuel appliance, if a room has a suspended wooden floor that is well ventilated, a good solution is to fit a simple floor vent to one or both sides of the hearth.

For a new construction using a solid ground floor the design should include two ducts from two walls at right angles to a mixing chamber below the floor and then from this to a grill in the floor.

Rooms with existing solid floors present a more difficult problem. Venting directly to the outside, preferably at a high level to minimize cold draughts, is a simple answer. However, be warned – a vent on the windward side of the house may introduce

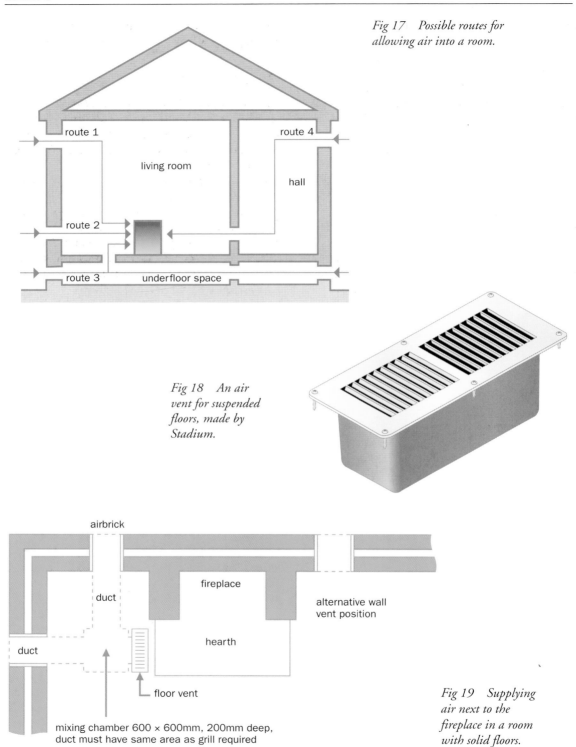

Fig 17   Possible routes for allowing air into a room.

route 1

route 4

living room

hall

route 2

route 3

underfloor space

Fig 18   An air vent for suspended floors, made by Stadium.

airbrick

duct

fireplace

alternative wall vent position

duct

hearth

floor vent

mixing chamber 600 × 600mm, 200mm deep, duct must have same area as grill required

Fig 19   Supplying air next to the fireplace in a room with solid floors.

excessive amounts of cold air. Alternatively, a vent on the lee side (low pressure side) of a house may actually suck air out of the room! This is explained in more detail in Chapter 12. This would tend to pull smoke and fumes back into the room, defeating the reason for the vent in the first place. Try to choose a wall in a neutral position in relation to the normally prevailing wind direction.

A better solution in this situation can be to vent from another room, conservatory or hallway and then to vent this room from outside. Such vents must have at least the same free open area as that laid down for the fire being served. If more than two vents are in series, then the subsequent ones must be increased by 50 per cent. Vents between rooms should be no higher than 450mm (18in) from floor level, to reduce the spread of smoke in the event of a house fire. An exception to this is the Draught Master vent, which has a self-closing flap that prevents a backward flow of gases and so can be fitted at a higher level.

EXTRACT FANS

If an open flue appliance is to be installed in a room then the installation of an extract fan should be avoided. In the case of solid fuel, the leading body, the Heating Equipment Testing and Approval Scheme (HETAS), states that a mechanical extract ventilation system must not be installed in the same room as a solid fuel appliance.

Where extract fans are fitted into a room in the property then additional ventilation may be required to prevent it affecting the room containing the appliance. This will avoid the extract fan creating a negative pressure in the property that may interfere with the operation of a chimney.

CHAPTER 3

Appliance Types and the Right Choice

There are many factors to take into account when choosing the correct appliance for your needs and circumstances. Some of these are basic choices such as the fuel type and whether you wish to use it to heat just one room or the whole house, but others are much more subtle and will be missed if you are not led through them carefully. These choices can be equally as important when ensuring end satisfaction with the appliance chosen.

Advice can be obtained from the various fuel suppliers and a list of contact numbers is given in the final section of this book, but these are not the only organizations able to provide advice and information. If you choose to employ a professional installer then they should be able to offer answers to any questions raised for any of the products they are offering; if not, then I would advise seeking an alternative installer.

Below is a basic description of the various types and their normal applications, but this is in no way exhaustive as there are so many products on the market it would be impossible to include them all in a book of this type.

I will start with the basic fuel types that are available.

FUEL CHOICES

Gas – Natural and Liquid Petroleum (LPG)

This is one of the fuels that most of us are familiar with as the vast majority of heating appliances currently installed in the UK are fuelled by this in some form or another.

It is interesting to note that the developments in gas-burning technology now allow some of the appliances to achieve over 95 per cent efficiency, but one of the popular types, the gas decorative fuel-effect open fire, is one of the least efficient types of appliance on the market. In some cases, these fires will provide almost no additional heat at all while burning large quantities of the world's premium fuel!

There are three types of gas in use in the UK and by far the most common is natural gas, which is the fuel delivered to us via the gas main and is consistent. These days, you can purchase your gas from a variety of suppliers, but they will all end up charging you for the same fuel, delivered through the same pipe with the same heat content and burning characteristics. The main choice here is price and service levels, but at least this enables you to use your purchasing power to obtain the services you require, which was not possible in the monopoly of the past.

Where a property is not connected to the gas mains it can prove very expensive to make that connection if the distance from the main is considerable, but even then all is not lost. It is still possible to get the convenience and efficiency of gas by having it delivered by tanker to the doorstep and stored in a high-pressure container. This is known as liquid petroleum gas or LPG. This is obtainable in two varieties and normally it will depend on the storage quantity and type as to which variety is chosen.

It is important to remember that this is not the same fuel as that delivered through the gas main and so all appliances have to be purchased with injectors and burners specially set up for the type of LPG you intend to burn.

Another consideration that is often forgotten is that this convenience comes at a price! With all the processing, delivery and storage costs LPG is considerably more expensive than natural gas and can in some areas be almost twice the cost per unit of heat. This is worth considering if it is an inefficient appliance you are thinking of installing.

Oil

There are a number of oil types and many are not suitable for domestic use, but there are two main grades used for domestic appliances. If you are intending to run more than one appliance from the same fuel store ensure they both use the same grade.

Oil appliances can also be over 95 per cent efficient and can result in the same level of control and efficiency as a gas appliance. It is unusual these days to get very inefficient oil appliances, but it is still important to consider efficiency when making your choice.

Because of the efficiency and controllability of appliances burning this fuel it is often considered the least expensive to use for central heating outside the reach of the gas mains. While this is essentially true, oil prices are still very volatile and can rise rapidly in times of great use or during regional conflicts in the Middle East, thus making running costs and availability variable. These factors must therefore be closely monitored to ensure that purchases of the fuel are made at the best possible time.

Solid Mineral Fuel

At one time this was the main heat source for UK homes and indeed we have to thank the original suppliers of this fuel for the development of the radiator-based central heating systems most of our homes enjoy today. With the increase in popularity of other fuels and the vast amounts of research and development money spent on appliances burning gas and oil, this fuel has fallen behind in the popularity stakes for main heating in our homes. However, the sale of solid fuel appliances, at least in multi-fuel guise, is once more gaining in popularity, but now as a focal point supplement to the existing central heating system in our homes.

Efficiencies again vary considerably from appliance to appliance, ranging from large open fires with a low or negative efficiency to appliances that can provide heat to more than one room at an efficiency of almost 80 per cent. It is at this stage impossible to manufacture appliances with efficiencies greater than this, due to the fact that the chemicals in the smoke condense out to form strong acids that will damage the chimney or flue as well as leaving insufficient energy in the gases to ensure they all evacuate through the chimney correctly.

There is a bewildering variety of solid mineral fuels, some of which are rated as smokeless and others that are not. Only smokeless fuels can be used in appliances installed in a smoke-control area even if the appliance is classed as a multi-fuel unit. (To find out if the property is in a smoke-control area, contact the local council environmental health department with the postcode of the property and they should be able to tell you.)

Solid fuel is often chosen for central heating for several reasons:

- it burns cleanly;
- it has high heat content for any given volume;
- it is more controllable in the way it burns;
- there is such a variety that if one does not suit then another is almost bound to.

Whether or not a solid fuel proves economical will depend on the appliance type, the fuel type, the way the appliance is used and the way the property is occupied.

If a property is left empty for large parts of the day and the heat is not required, then the constant burning associated with solid mineral fuels may not be economical. In houses where someone works from home or the house is occupied for long periods by young children or an elderly relative, then solid fuel can be very competitive with other fuels and may even turn out to be cheaper than using gas or oil almost constantly.

New developments in solid fuel central heating such as pump-assisted systems, link-up (two boilers into the same heating system) and thermal store technology are all improving the controllability and efficiency of solid fuel heating systems.

Wood Burning – Including Wood Pellet Burners

This is now recognized as the most environmentally friendly form of heating based on combustion. The amount of carbon dioxide produced when burning a mature tree is less then it would have absorbed during the growing process. This means that if the wood fuel is taken from managed forests where the felled tree is replaced, then the use of wood is CO_2 negative. In addition to this, the burning of a log produces less ozone-damaging gases than allowing the log to rot on the ground, plus of course this is a renewable energy source in that gas, oil and coal take millions of years to form, but a new tree can reach a sufficient size for cropping within fifteen to twenty years.

Since it is so environmentally friendly why is wood not a more widely used heating fuel? There are many limiting factors and problems associated with burning wood; some are unavoidable, while others are purely as a result of poor user operation and maintenance. Some of these are given below.

- The calorific value of the various kinds of wood will vary considerably, but will always be less than other solid fuels. (In the case of pine there will be only a quarter of the heat value per kilogram than that of a good smokeless fuel.)
- The bulk density (the weight of one cubic metre of fuel) can be as little as a quarter of that of a good smokeless fuel.
- Some timbers, such as the ones with better calorific values and greater bulk densities, can take up to three years' preparation before being ready to burn.
- The above factors mean that a much larger space is required for storage than with any other fuel.
- The low calorific value and bulk density mean that the appliance will need refuelling anything from three to eight times more often than when using a solid mineral fuel to achieve the same heat output for the same amount of time.
- The burning process will often cause tar formation on the heat transfer surfaces of a water boiler and so wood is not ideally suited to the most common form of central heating used in the UK.
- The positioning of the chimney and appliance

and the use of warm air distribution are critical for whole-house heating from wood.

Many of the traditional perceived failings of solid fuel can be eliminated when burning wood pellets and this is considered to be the future for solid fuel burning and wood burning in particular. Modern wood pellet stoves have a hopper feed for the fuel (allowing much more fuel to be loaded at one time), electric ignition (so the fire can be allowed to go out when not required and automatically relit when needed again), low ash production and full thermostatic control. At the time of writing the government was sponsoring the Clear Skies initiative by offering grants for the installation of some wood-burning appliances.

Electric

Despite the perception that electric heating is environmentally friendly and efficient, this is not in fact the case. In the first instance, electricity has to be generated and even today the vast majority of this generation is done via the burning of fossil fuels, that is, gas, oil or coal, and this is done at relatively low efficiencies, normally between 35–45 per cent. When transported through the grid there is a further 3–5 per cent loss, and then when used in the home the appliance can be between 90–100 per cent efficient. This results in a prime fuel being used at less than 40 per cent efficiency when the electricity is used in the home. In addition, the pollution caused by electricity generation still exists around the environment of the power plant.

If the above factors are taken into consideration electricity becomes the least efficient form of heating and is far from environmentally friendly (although this situation is improving all the time with the use of renewable energy sources for electricity generation, although less than 10 per cent of UK electricity is generated in this way).

Electricity has its advantages nonetheless, including cleanliness, rapid response (for direct radiant-type heaters), ease of operation, no fuel storage required and the fact that most houses in the UK possess an electrical supply, even if some have to generate their own. One of the most important factors in favour of electric heating is that there is no

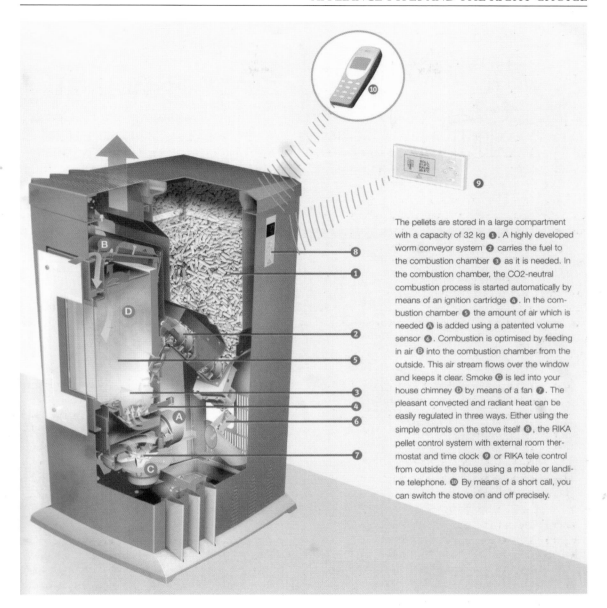

The pellets are stored in a large compartment with a capacity of 32 kg ❶. A highly developed worm conveyor system ❷ carries the fuel to the combustion chamber ❸ as it is needed. In the combustion chamber, the CO_2-neutral combustion process is started automatically by means of an ignition cartridge ❹. In the combustion chamber ❺ the amount of air which is needed Ⓐ is added using a patented volume sensor ❻. Combustion is optimised by feeding in air Ⓑ into the combustion chamber from the outside. This air stream flows over the window and keeps it clear. Smoke Ⓒ is led into your house chimney Ⓓ by means of a fan ❼. The pleasant convected and radiant heat can be easily regulated in three ways. Either using the simple controls on the stove itself ❽, the RIKA pellet control system with external room thermostat and time clock ❾ or RIKA tele control from outside the house using a mobile or landline telephone. ❿ By means of a short call, you can switch the stove on and off precisely.

Fig 20   A sectional diagram of a Rika wood-pellet-burning stove.

need for a chimney or flue, which will often make the capital cost (initial cost of installation) lower than other forms of heating, although this has to be set against the relatively high running costs (compared with gas and oil).

APPLIANCE TYPES

Some types of appliance have a 'decorative effect' and have the heat provided by a separate heat source. An example of this is a multi-fuel stove-effect electric heater. In this case, there is a visual effect behind a

29

Fig 21   A sectional diagram of a Nestor Martin Stanford log-burning stove.

glass door but the heat is provided by an electric fan heater. Below is a description of most of the appliance types.

Central Heating Boiler

This is the most common type of heating appliance in the UK and the vast majority use gas as a fuel (either natural gas or LPG), but are available with all the fuel types discussed above.

The principle of this type of heating is to heat water through a heat exchanger and when this has reached the desired temperature distribute it around the house via pipes and radiators. The advantage here is that the one appliance can be used to heat the whole of the house, so that it is often the most cost-effective way of heating a home as far as capital cost is concerned.

Due to the market size, manufacturers have spent vast sums on research and development, particularly in the gas and oil sectors, to develop new and better models. With the push for ever greater efficiency and control these units are now available with efficiencies of 95 per cent or more. This type of condensing boiler recovers heat from the condensing process that takes place in the flue gases by making the condensation occur in the heat exchanger. These units are

Fig 22   A section through a balanced-flue gas fire.

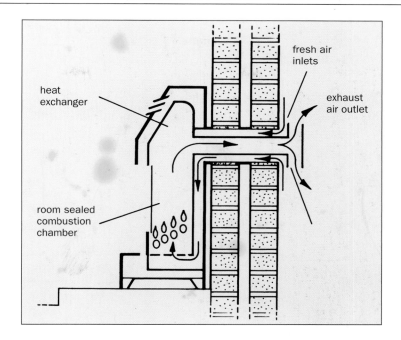

heat exchanger

fresh air inlets

exhaust air outlet

room sealed combustion chamber

complicated and require specialist knowledge to ensure safe and effective installation. In addition, the increased degree of development and technological advance have resulted in high capital costs, reduced reliability and greater repair fees when things go wrong. Some traditionalists still believe that the increase in capital and repair costs outweighs the efficiency benefits and so they remain with the 80–85 per cent efficiency boilers.

Many central heating units use fan or balanced flues that come as a part of the appliance and pass directly out of the wall. A balanced flue will also have an inlet for the air needed for combustion to enter the burner area direct from outside. This type of unit avoids the need for a chimney or vertical flue and eliminates the potential draughts caused by bringing air into the room containing the appliance.

It is possible to get electric boilers for use with water-based central heating systems and these contain a high-mass heat store that is heated via electric elements using lower cost electricity, during agreed low-rate periods, and then the heat is released over a period of time into the water passing through the central heating system. To ensure storage of sufficient heat these are often large units of great weight and cost.

Other types of units are available, including thermal store boilers and combination boilers, both of which provide hot water at the same pressure as the cold water. This has advantages for showers and for ease in locating the boilers and feed tanks.

As well as simplifying the heating system layout a combination boiler avoids the need for storing hot water and so reduces the heat loss from the hot water side of the system. A modern combination boiler will provide sufficient hot water for showers, baths and a number of sink units, but the boiler is again increased in complication, and when combined with a condensing heat exchanger can become so complicated that fault diagnosis is beyond many heating engineers.

The appliances listed above will not require a chimney hearth or fireplace and so these will no longer be considered in this book, but the appliances described below will all require one or other of these and so are the types of appliances that will be considered from now on.

There are still some gas and oil appliances that require a vertical flue, but these are becoming rare. However, in the case of solid fuels, both mineral and wood, there is still the need for a chimney or vertical flue. There have been a number of balanced flue

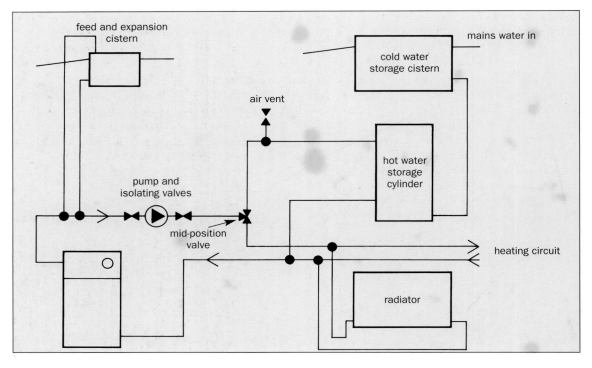

Fig 23   A heating system with stored hot water and a standard boiler.

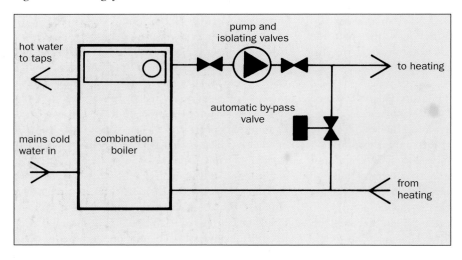

Fig 24   A system using a combination boiler with an instantaneous hot water system.

wood-pellet-burning appliances being imported, but these are still rare and their compliance with the Building Regulations and the Clean Air Act still has to be resolved.

The current Part L of the Building Regulations requires solid fuel appliances that provide central heating to have a water-temperature-sensing thermostat and to have an efficiency that has been independently verified as being at least equal to that required to achieve HETAS approval. Solid fuel appliances are not as popular as they used to be because they require the user to deliver the fuel to the appliance by hand as well as clean out the ash deposits and keep the flue ways clean. Despite advances in design making these units easier to use and clean, there is still an element of work required.

Solid fuel appliances have a category called a solid fuel room heater with boiler. This is still the most popular of the solid fuel central heating appliances. It is, in fact, a room heater or stove that has an integral high-output boiler that is controlled by a water-temperature-sensing thermostat built into the unit. It is designed to provide heat into the room in which it stands as well as heat the water for the central heating and the domestic hot water. This type of central heating unit is probably the simplest in design and

construction and therefore is very reliable and easy to operate. Many of these units will last for fifteen to twenty years without anything other than normal user operation and annual servicing (which must include a chimney sweeping).

The design of the central heating systems connected to these appliances is specialized and some further information on this must be sought, but system design has advanced sufficiently to enable a higher than expected degree of controllability and efficiency to be achieved.

Room Heaters and Stoves

In many cases these are again split into categories,

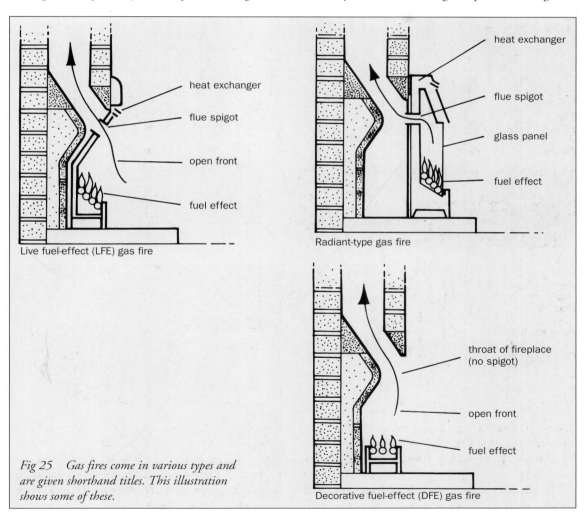

Fig 25   Gas fires come in various types and are given shorthand titles. This illustration shows some of these.

especially in the case of gas. In the past, the most common type of gas room heater would have been the radiant gas fire – an example of this most people my age well remember is the Cannon Gas Miser! The radiant gas heater has been around for many years and was the original form of gas heating unit. The gas fire has come a long way since then and in many cases is designed to be more decorative. The most common types are those that have some type of imitation solid fuel fire, sometimes open and sometimes with a glass panel in front (I exclude from these the category of decorative fuel-effect gas fires that are considered under the section on open fires). Figure 25 shows the different types of gas fires in order to give an indication of the descriptions given, so the reader can identify the types from the shorthand titles used by heating engineers.

Fig 26   ABOVE: *An illustration of a live fuel-effect gas fire.*

Fig 27   TOP RIGHT: *An example of a decorative fuel-effect gas fire.*

Fig 28   *An inset studio-style gas fire.*

Fig 29   An electric room heater with a fan heater and flame effect.

There are also electric fires that would come under the category of room heaters and these are available as direct radiant heaters and fan heaters. Although it is possible for these to be portable white-cased units, this is not the type discussed here. The types under consideration are those that form a focal point to the room and often seek to imitate a solid fuel fire. Their success is in the eye of the beholder, but the effects are improving all the time. These fires do not require a hearth, fireplace, flue or chimney, but they are often positioned in front of a decorative surround for effect.

The oil industry has entered this market sector in a big way over the last decade and has made great improvements to its products. Recently, the oil market has also sought to imitate the solid fuel fires, and appliance manufacturers are producing oil-fired stoves in some numbers.

This all leads us to the original of the genre, the solid fuel stove. In this section, as in the open fire section, this is what all the other fuel types strive to emulate, in some cases rather well, although none has succeeded completely in capturing the ever-changing, almost living nature of a solid fuel fire.

Solid fuel appliances, whether burning solid mineral fuels or wood-based fuels, all need a hearth and a chimney, and in many cases a fireplace. These stoves vary in quality of construction, design and appearance, but it is almost certain that one can be found to suit each individual's property and requirements.

What must be borne in mind is that appearance must not be the only criterion used when making a choice. For example, if a stove is too small to heat the room it may have to be worked so hard that it will be damaged, but on the contrary one that is far too large in heat output may result in damage to the chimney and a dangerous installation.

Inset Room Heaters and Stoves

This type of stove is similar to the above type, but instead of standing alone inside a fireplace or in front of a chimney it is built into the chimney breast.

This can often save space and allow the use of a convection chamber to create a more even distribution of hot air around the room, which will help to avoid the hot spot around the appliance and the cold spots in the corners of the room. Some of these appliances even have built-in convection chambers that allow the connection of ducting to facilitate the distribution of warm air, either by natural or fan-assisted convection, to other rooms in the property.

Fig 30 An oil fired coal-effect room heater.

Fig 31 An Esse Dolphin stove in a Victorian style.

One such appliance is the Nestor Martin inset shown in Figure 35, but many of this type of unit are available powered by solid fuel, gas or oil and so the choice becomes even greater. This type of inset appliance is also popular for solid fuel appliances with high-output boilers for central heating. The installation method allows insulation to be used around the boiler, increasing the proportion of heat retained in the boiler itself.

As can be seen from the photographs, the style of solid fuel inset stoves and room heaters varies and again most tastes can be catered for, but should this type of appliance be required to run the central heating it is even more important to choose the correct size.

OPPOSITE: Fig 32 A Charnwood country stove – traditional design combined with efficiency.

Fig 33   A Villager Puffin contemporary style with small output.

Fig 34   A Charnwood L/A inset room heater with central heating boiler.

Fig 35   A Nestor Martin convection inset with ducted warm air.

Fig 36   An ESSE Furnesse inset convector which can be fitted directly into an existing open fireplace.

Fig 37   A Jetmaster convector open fire in an inglenook fireplace.

Open Fires

When we think of fireplaces and chimneys this is often the first thought we have, and in fact it is still one of the most popular forms of fire today. An open fire instantly adds that extra something that gives a cosy feel and indefinable 'feel-good factor'. The only fuel choices for this type of fire are solid fuels or gas. The convenience of the gas-fired open fire has long since held an appeal that has resulted in many thousands being installed, but even today they cannot quite match the original feeling of a real fire. This is the type of fire that most satisfies the primeval need to see a fire and to feel the radiant warmth they generate.

Even in this most basic of systems there are levels of sophistication and design that will enable open fires to operate at anything from little, or no, efficiency up to 50 per cent efficiency.

The Jetmaster fire shown here is a wood-burning open fire that is capable of providing 12kW of heat at

Fig 38 A Dovré 2000 open fire with folding doors to close off for greater efficiency.

almost 50 per cent efficiency while a simple log basket in a large opening is likely to produce less than 3kW of heat at less than 5 per cent efficiency.

If deciding between an open fire and a stove becomes difficult there are always closable open fires or openable stoves! The Dovré 2000 (Figure 38) can operate at 45 per cent efficiency as an open fire, but if you wish the foldaway doors can be closed to create a stove operating at nearly 80 per cent efficiency.

Also shown here is the Stovax combination convector fire that looks like a reproduction Victorian tiled register grate, but has a convection chamber incorporated to improve heat output and efficiency.

Fig 39 A Stovax reproduction fireplace with built-in convection chamber.

Fig 40   A Parkray Cumbria inset room heater with a boiler and traditional styling.

CHAPTER 4

Chimneys

In the UK there is a mystery surrounding chimneys which is based on myth and legend rather than science and fact. This has resulted in a lack of knowledge both in the building and architectural fields which is rapidly spreading to the specialist installer section of the industry.

It appears that many people are willing to install modern appliances into flues that were constructed tens, or even hundreds, of years ago, and worse still are willing to construct new chimneys based on the old wives' tales that were around at the turn of the century. Examples of these include:

- all flues must have a bend in them to increase the draw;
- a standard 225 × 225mm (9¾ × 9¾in) chimney is suitable for all types of appliances;
- a bit of smoke now and then never hurt anyone.

All of the above are incorrect and need to be dispelled before we can consider the good design of a chimney. However, before an understanding of what is a good design can be achieved it is necessary to have a basic understanding of how a chimney works, because only then can we see what will have a beneficial effect and what will hinder the correct operation of the chimney.

HOW A CHIMNEY WORKS

A flue works as a result of two natural principles. These are pressure difference and temperature difference.

Pressure Difference

It is a well known fact that the higher above sea level we go the lower the atmospheric pressure. This reduction in pressure occurs because there is less air above us and so less weight. To understand this, we have to accept that air has a weight just as water or any other substance has. The weight of air may be

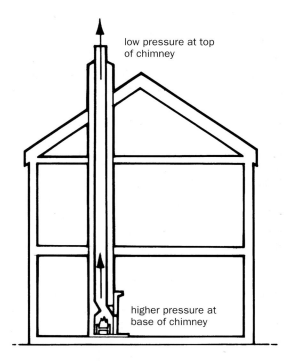

low pressure at top of chimney

higher pressure at base of chimney

Fig 41   Air moves from a high-pressure region to a low-pressure region, creating the initial draw in a flue.

44

Fig 42   Air has a weight – the warmer it is, the less that weight is. Cold air will therefore force down, while warm air will rise.

many times less than the other substances we are used to lifting, but it is still there.

The height difference between the base of the flue (the top of any appliance) and the terminal of the flue will cause a pressure difference, because at the top of the flue there is less air above us than when we are at the bottom of the flue. If the low-pressure and high-pressure regions are directly connected, as in this case by the flue within the chimney, the air will travel from a high-pressure region, at the base, to a low-pressure region, at the top, in an attempt to balance out the difference. It is this movement of air from the base to the top of the flue that results in the initial draw of a chimney. This initial draw has to be present or it would be impossible to light the fire in the grate without all the smoke entering the room.

Temperature Difference

Warm air is less dense than cold air and so any given volume is lighter. This means the gases in the flue are lighter than an equivalent column of gases in the ambient air. This difference in weight means the hot air rises and adds to the static draw of the flue. Because a flue for a solid fuel appliance has to draw oxygen into the room, through the appliance and firebed as well as drawing the products up the

chimney, it has to be kept warmer to maintain as much of the temperature difference as possible.

Once the fire has been lit, the air in the flue will begin to get hotter, and the hotter the air in the chimney the lighter it becomes. As the temperature difference between the air in the flue and the air outside begins to increase, then the faster the hot air will travel up the flue and the greater the draw will become. Eventually the flue will reach its maximum temperature and the draw will reach its final optimum amount.

This gives us the main positive influences on the effectiveness of a chimney: firstly, the taller the chimney the greater its initial draw; and, secondly, the warmer the chimney the better the increase in draw when the fire is lit at the bottom of the flue, and so the best chimneys are tall and warm. This, of course, will depend on other factors – for example, a very tall chimney may end up being very cold and as a result will not improve in draw once the fire has been lit.

On the other hand, it may be possible to create a short chimney that is well insulated and protected, so that it will be slow to start but will increase in draw dramatically and quickly once the fire has been lit. The skill of the chimney designer is to ensure

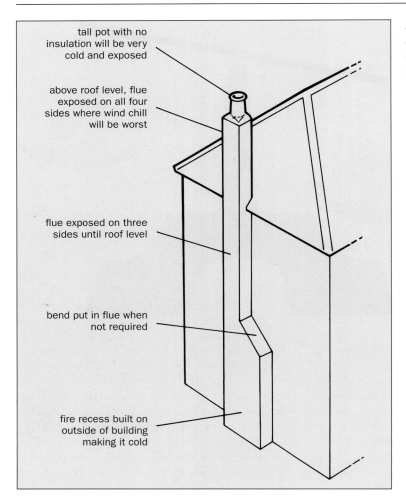

tall pot with no insulation will be very cold and exposed

above roof level, flue exposed on all four sides where wind chill will be worst

flue exposed on three sides until roof level

bend put in flue when not required

fire recess built on outside of building making it cold

Fig 43 If a chimney is external and has a tall stack it may be exposed and cold, making it inefficient.

adequate draw initially and enable this draw to increase significantly when the fire is in use.

CHIMNEY ENERGY LOSSES

As with all such energy systems, there are not only gains to be had but also losses to consider. Above we have shown the gains in a chimney system, so we must now consider some of the losses as it is only by striking a balance between the two that we can provide a chimney that is suitable for the appliance we intend it to serve. The losses come in two main categories – heat losses and frictional losses. The first section is the easiest to understand, estimate or calculate, so this will be the section we will start with.

Heat Losses

The first thing to consider is how these heat losses occur: one is the reduction in the heat input to the system; the other is the loss from the structure. Let's take a look at the reduced input first. This is the easiest to grasp, but is often not given sufficient consideration when the appliance using a chimney is initially chosen or changed.

The size of an appliance, its type and its efficiency will all result in a given heat input into the chimney. If it is a large appliance that has a large heat output it is likely to give a greater heat input into the chimney than a small appliance. But even this is not as simple as it may at first appear. Let's look at a series of examples below:

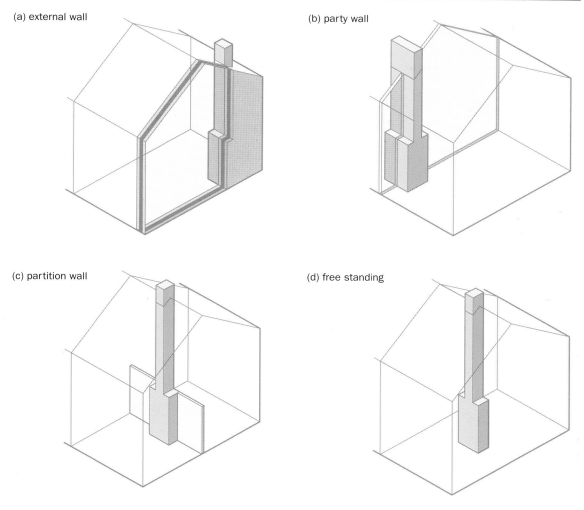

(a) external wall

(b) party wall

(c) partition wall

(d) free standing

Fig 44   The location of a flue will affect how warm it will remain: (a) It will be coldest on an external wall. (b) Multi-flue chimneys are cold if only one flue is in use; the tall stack from below the ridge will also cool down. (c) Through the ridge is better, as is a single flue. (d) Free-standing through the ridge is best because it only has to warm itself, not any brick walls around it.

- An open fire using a dog basket in a fireplace with an opening of 600 × 600mm (23 × 23in) will have an efficiency of approximately 15 per cent and will give only 2.7kW to the room: (2.7/15) × 85 = 15.3kW up the chimney.
- An open fire installed using British Standard components and installed in accordance with the relevant British Standard will operate at approximately 28 per cent efficiency and give 3.0kW

to the room: (3.0/28) × 72 = 7.7kW up the chimney.

With any open fire, the efficiency is low and so the amount of heat entering the chimney to keep it warm will be quite high, but when we start to look at the equation for closed appliances this can be a very different story depending on their individual output and efficiency:

- A large stove of 10kW output working at an efficiency of 60 per cent will put the following heat into the chimney when working at its maximum burning rate: (10/60) × 40 = 6.6kW heat input to the chimney.
- A small solid fuel stove with a maximum output of around 4kW and an efficiency of 60 per cent will provide a smaller amount of heat into the chimney when working hard: (4.00/60) × 40 = 2.66kW heat input to the chimney.
- A large high-efficiency stove operating with an output of 10kW and an efficiency of 80 per cent will give the following result when working hard: (10/80) × 20 = 2.5kW heat input to the chimney.

We will now look at one of the most common situations in energy-efficient properties where a small high-efficiency stove is used:

- With a small high-efficiency stove with an output of 5.0kW and an efficiency of 80 per cent it can be seen that the amount of heat entering the flue is very low: (5.0/80) × 20 = 1.25kW heat input to the chimney.

It must also be understood that a small high-efficiency stove is not likely to be used at its full output for much of the time, and that due to the control over a closed appliance it can be turned down to an output of as little as 1.0kW to the room. When an appliance is turned down to a lower setting it will not operate at its best efficiency, but a good quality stove may be capable of operating at 70 per cent efficiency when turned to its lowest output. If we take this into account we can see that the amount of heat entering the flue is minimal in this case: (1.0/70) × 30 = 0.43kW heat input to the chimney.

It can be seen from the examples above that it would be wrong to expect the same chimney to operate effectively in all cases. If a chimney is designed to operate with a dog basket in a large opening it will not need to have the same level of heat retention as for a small high-efficiency stove. The volume of the flue will have to be much greater for the open fire to be able to cope with the amount of air that will pass through it, but in the case of the small stove this large volume could not be kept warm.

The above is one of the fundamental considerations when looking at how to design a chimney, but the next section is of equal importance. When we design a chimney we not only have to take into account how much heat will be entering the flue, but also how much heat the flue will lose. The heat loss from a flue will depend on many things and listed below are the most important ones, which will then be explained in further detail.

Heat loss is measured in units of watts per square metre per degrees Centigrade (W/m^2°C). From this, we can start to unravel the important factors.

Firstly, the area of the flue's internal surface will make a difference. For instance, let us consider a straight, round flue of 225mm (9¾in) internal diameter and an effective flue height of 6m (19¾ft), then compare this with a flue of the same height but with an internal diameter of only 150mm (6in) – there will be a reduction in area of almost 35 per cent. The comparison is highlighted in the feature box.

Differences in Internal Flue Surface Area with Flue Diameter

- Flue internal diameter = 225mm or 0.225m.
- The perimeter or circumference of the flue will be the diameter × π (this symbol represents the constant known as pie and is calculated as 22/7).
- The surface area of the flue will be the perimeter × the height and so the equation for the surface area of a round flue will be: (internal diameter × 22/7) × the effective flue height.
- In this example, this results in the sum: (0.225 × 22/7) × 6 = 4.24m^2.
- If the same equation is used for the 150mm internal diameter flue then the following sum results: (0.150 × 22/7) × 6 = 2.82m^2.

This simple example shows that if all other factors are the same, the 150mm (6in) diameter flue has only 66 per cent of the surface area of a 225mm (9¾in) diameter flue, and so will have only 66 per cent of the heat loss of the larger flue.

However, the calculations shown here are not the end of the story, as the temperature difference between the inside of the flue and the outside will also make a difference. If the temperature in the flue is an average of 120°C and the outside of the flue is an average of 20°C, as would be the case if nearly all of the flue passed inside the house, then the temperature difference would be 100°C. If the flue passed up the outside wall of the house with three sides exposed to the cold outside air, then the average external temperature could be as low as –1°C overnight in the winter. In this case, the temperature difference would be as high as 121°C and so the heat loss would be 21 per cent greater than for the internal flue.

These are very basic considerations and can be observed when looking at a chimney without any specialist knowledge, but there are other factors that will need to be considered with regards to heat loss.

In the newest version of the Building Regulations it states that the space between any flue liner and the brick wall surrounding it must be filled with an insulating material to reduce the heat loss from the flue, but in older flues (those constructed before April 2002), this material may not be insulating. As we all know, the amount of heat conducted by some materials is greater than others. If we have a pan on the cooker and it has a metal handle we will expect this to be hotter than if it has a wooden handle, or if we place a hot drink inside a metal mug it will cool quicker than if it is placed in a thermos flask. Both of these are examples of how heat will travel through materials at different rates.

In the case of the thermos flask, this does not consist of a single material but a combination of materials that come together to form an insulating structure, and this is what our aim should be when constructing a chimney. It is necessary to have a solid, rigid material to provide the structure with strength and integrity, but this type of material is not normally very insulating. It is also necessary to have a solid, heat-resistant material to form the flue liner and some of these are insulating while others are not. To provide the maximum heat retention, an amount of heat-retaining material should be included between the above two components. This heat-retaining material, and the amount of it, becomes more important when the surface area of the flue is large or where the amount of heat being input to the flue could be low. The structure and materials used in a chimney will also affect its ability to cope with whatever appliance is installed at its base.

Frictional Resistance

Having studied the negative influence of heat loss on a chimney's ability to function efficiently, we need now to look at the additional negative influences that are caused by the various frictional losses within a flue. There are many influences on friction levels within a flue, but here we will deal with the most important.

The surface of the liner is going to be the biggest influence in most flues; this is with regard to both amount and type. If we again consider the internal surface area of a flue it will become evident that the larger the surface area, the greater the amount of friction that will be caused, although this is a simplified version of the facts as will become apparent later.

The other issue is, of course, how easily the smoke is able to pass over the surface – as with all fluids and solids, the smoother the surface the easier the smoke will pass over it. It can be difficult to appreciate the fact that a gas will have a resistance due to friction when it passes over a solid (in our case the smoke passing over the surface of the liner). Consider as an example the wind blowing through a tree. The stronger the wind, the more the tree will bend or deform; also the more leaves on the tree, the greater the extent to which the tree will bend. The force causing the bending is the resistance that the surface of the tree presents to the wind. As we consider this, we can see that if we keep the surface area of the flue to a minimum and the surface finish as smooth as possible, we will reduce the amount of resistance caused.

Other items that cause resistance in flues are bends and changes in size. Consider the amount of energy a car uses when passing along a road. If the road is straight and stays as a dual carriageway so that a constant speed can be maintained, then we are able to achieve a fuel consumption of, say, 7ltr/100km (40mpg). If we travel the same distance down a road with a number of sharp bends and changes from single track road to dual carriageway we will have to change speed often and the extra resistance of the tyres when going around the bends will increase friction, and as a result we will be lucky to achieve 9.4ltr/100km (30mpg).

It follows, therefore, that the least resistance is created by a straight flue that does not change its size at any point, not even at the chimney pot. In the case of open fires the next most important issue is that of the gather at the base of the flue, where all the gases are compressed into the flue from the larger fireplace opening.

It is a matter of scientific fact that if a fluid, gas or liquid, is forced from a large duct into a small duct, then the fluid will have to speed up to be able to pass through. We can see this when we look at a weir on a river. The water passing down the river on the upper side of the weir appears to be flowing very slowly as the river is wide and deep at that point, but when the water is forced to pass over the weir it is now very shallow and so it has to speed up considerably.

The same applies to the air and smoke in our fireplace. At the fireplace opening the area for the gases to pass through is quite large, for example for a 600 × 600mm (23 × 23in) fireplace the area is 0.36m² (3.7sq m). It then passes through the gather and enters the flue, which has an internal diameter of 225mm (9¾in) and an area of only 0.05m² (0.5sq ft). To achieve this, the gases have to speed up to over seven times their original speed.

If we expect the gases to exit instantly through a flat top with a hole in the middle, the resistance will be very high and it will take an enormous amount of energy. If, on the other hand, we make this happen gradually by forming a gather, then this will cause less resistance and take less energy. To be able to visualize this, consider a flat plate with a hole in the centre and pour water onto the plate – it is almost inevitable that some of the water will spill over the edge of the plate; similarly, smoke will spill from our flat-topped fireplace into the room. If, however, we take a funnel and repeat the same process, unless we poor the water into the funnel too fast, we will be able to capture all the water; similarly, if we make a gradual gather we are more likely to be able to catch all the smoke. The better the angle of the gather and the smoother the surface of this gather, the less resistance it will cause to the smoke entering the flue.

This explanation leads us on to the next issue for an open fire, which is the volume of gases entering the flue and the speed at which they will have to travel in order to all pass up the flue without entering the room. The air passing into the fireplace from the room needs to do so at a given speed to be able to catch all the smoke and drag it into the flue; this is taken as approximately 0.3 metres per second (m/s). When the gases are compressed by the gather to direct them into the flue, they are speeded up so that the volume of gases can all travel up the flue. The greater the change in area from the fireplace area to the flue area the more the gases will have to be speeded up. The faster the gases travel up the chimney, the more resistance they cause and so the more energy they require.

To make another analogy, we consider our car on the road and the fuel consumption again. If we travel on a motorway at 80km/h (50mph), then the fuel consumption will be a lot less than if we travel at the national speed limit of 112km/h (70mph). The increase in fuel consumption is due mainly to the extra frictional resistance caused by the air travelling over the car faster. This increase in fuel consumption will happen even if the car is streamlined, but will be less than if it is not. The same applies in our chimney – the faster the flue gases pass through the flue, the greater the increase in resistance. This increase will be less if we have a smooth, straight flue with a gradual gather than if we have a rough surface with bends and a sharp change in shape or area, but it will still increase.

This shows that if we are to get all the gases to pass up the flue, we will have to control the increase in speed to match the amount of positive energy being created or otherwise smoke will enter the room. To achieve this will require a very careful consideration of all the factors above and the completion of a series of complicated calculations and reference to a number of tables for resistance factors and so on.

To overcome this and help with the design of chimneys, various graphs are available to enable size of flue and size of fireplace to be matched, taking into account 'normal materials and construction techniques'. It is important to realize that should you choose to stray from this 'norm', then the risks of failure are great, and obtaining a set of calculations

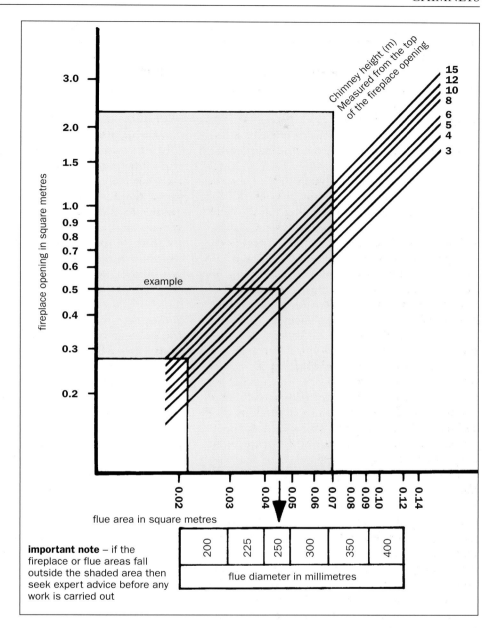

Fig 45
Graph for the
sizing of flues
for open fires
based on the
Scottish
Building
Regulations.

from a chimney specialist is important. Shown above is the graph given in the Scottish Building Regulations section F, and it is as good as any available. Another source of this type of information is the Solid Fuel Association (*see* Useful Addresses at the end of this book).

OTHER DESIGN CONSIDERATIONS

It is important to understand and accept the principles above, especially when attempting to assess an existing chimney, but it is also important to realize that for a chimney to be safe and well designed it has to provide functions other than just getting the smoke, or fumes, out of the room and up into the atmosphere. Below is a brief discussion of each additional function and the effects this may have on the design of the chimney.

Structural Stability

A chimney is of no use to a fireplace, property or user if it is unstable. It may take all the smoke from the room when first constructed, but will not do so for long unless it is designed and built to last for many years.

This subject is not one we can go into in great detail here, but it must be considered seriously. The structure must be designed so that it meets the requirements of the Building Regulations and any foundations or other supports must also be fit for their purpose. The main issue here is the stability of the chimney where it is unsupported above the roof. The section on Building Regulations in Chapter 5 covers this issue.

Fire Protection for the Building

A chimney has to be designed to contain the hot gases and protect the property from the effects of heat. This is most obvious at the base of the flue where a chamber is constructed to house the combustion appliance. This is known as the fireplace recess (previously called the builder's opening), and must not be confused with the fireplace opening, which is the final size of the fireplace.

At this point, we need to have a structure that will contain the heat from a fire contained in nothing more than a log basket or a coal grate. The radiant heat from these can be considerable and so this structure has to be constructed of non-combustible materials and have sufficient thickness to avoid any combustible material placed on or near the basket or grate overheating or catching fire. Where a stove or room heater is fitted, the flue pipe will become very hot when used at maximum output and so again only non-combustible material can be placed close to it.

In the chimney itself the temperature of the flue gases will be far lower, especially higher up the flue, which leads some people to underestimate the potential dangers at this point. It is a fact that the flue gases are unlikely to reach higher than 350–400°C under normal operation and will be below 250°C where an open fire has been fitted, but this is not the only condition that the fire resistance of a chimney has to be designed to cope with. In the event of a chimney fire starting low down in a flue, which is the most common location for one to start, then the whole length of the flue will be subject to very high temperatures, possibly even the flue pipe itself if one is fitted.

The temperatures in a flue during a chimney fire will depend on many things, such as the amount and type of deposits present and whether, or not, the air entering the flue can be contained. If a fire starts in a chimney where coal soot is deposited in small quantities, then it may reach temperatures of 800–1,000°C for a time of less than five minutes, but if there is a large amount of tar from the poor burning of wet wood, then the fire could reach temperatures of over 1,100°C for twenty to thirty minutes. A great deal of damage may occur, and this must be contained to damage within the flue if the house is not to burn down. Fire protection details are also covered in the Building Regulations section J in England and Wales and section F in Scotland.

Evacuation of Gases to Atmosphere

This may initially appear to be the same as the task described above in 'Frictional Resistance', but this section extends the process to keeping the gases contained as they pass through the rest of the property and out to the atmosphere.

Once the gases have entered the flue they must not be allowed to leak out into the house at any point. This is one of the most ignored elements of a chimney while it is being constructed, as there is a mistaken belief that if a chimney leaks it will draw air into the flue and not let the gases out.

It is a fact that under normal operating conditions the chimney will create a suction, and it is this suction that draws the smoke into the flue in the first place. It has to be understood that as the gases pass up the chimney and cool down they become heavier than those gases below. This extra weight makes the cooler gases at the top of the flue slow down and so the hotter gases below have to force them out of the top of the chimney. It is this variance in weight and force within a chimney, should it become great, which can force the hotter gases to take an easier route out of the flue rather than push the colder gases out of the top. If there is a leak in the flue at this point the gases can be forced out of this leak and into the property.

This may seem like an extreme case, and indeed it does require a significant level of leakage or a significant cooling of the gases, but the only way to be sure of avoiding its occurrence is to construct the flue in such a way that is as sealed and leak-proof as possible. The Building Regulations and British Standards not only provide advice on how to do this, but also on how to test for leakage in a completed or existing chimney.

Provision for Care and Maintenance

When a flue is new and clean it is at its most effective, just like any new item, but over time there will be the deposits of soot and tar when burning solid fuels, or dust and cobwebs when burning gas. These deposits will eventually build up to a point that they restrict the movement of gases up the flue and may eventually cause the spillage of gases into the room, or will become damp and start to attach to the walls of the flue as acids or alkalis. This can be avoided with regular maintenance and so it is important to recognize the amount and type of maintenance needed for the type of appliance to be installed and to provide adequate access to ensure this can be carried out.

Even where adequate maintenance is carried out, there will still be degradation over the years and so it is important that adequate provision for inspection and maintenance is made when designing and constructing a chimney. Unnoticed or unchecked degradation will make the flue and chimney unsafe over the years. Again, this element of the design and construction is covered in the Building Regulations and British Standards.

As shown above, there are regulations to comply with as well as recognized good practice standards to follow. A lot of people seem to think of these as red tape and restrictive in their application, but in fact they are a very good way of ensuring that what you design and construct will meet the needs of the appliance, property and user, as well as providing a safe chimney system.

In Chapter 5 we will therefore look at the regulations that need to be followed and then in Chapter 6 we will look at the application of these for each chimney type.

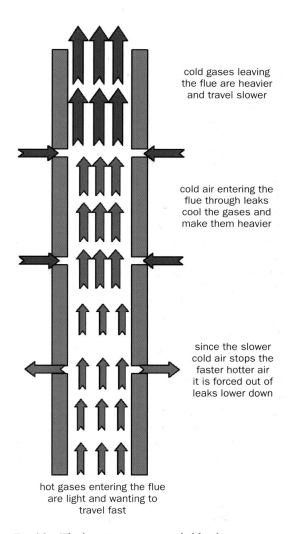

cold gases leaving the flue are heavier and travel slower

cold air entering the flue through leaks cool the gases and make them heavier

since the slower cold air stops the faster hotter air it is forced out of leaks lower down

hot gases entering the flue are light and wanting to travel fast

Fig 46   The heavier gases are cooled by the leaks into the flue, which will cause a pressure lower in the chimney and in the extreme case can force fumes out of the chimney from lower down.

CHAPTER 5

Regulations and Standards

Here we will look at the various types of regulations and where and when they apply. It is important to realize that only general guidance can be given in this chapter, so if you have any doubts or questions about your specific case you must contact the relevant body.

Each of the items below can be applied to some or all of the chimneys being built, brought back into use or refurbished, but even if some are not mandatory for the work you wish to carry out it is often sensible to follow the information and advice they provide.

PLANNING REGULATIONS

General

Planning Regulations are intended to protect the environment in which we live and so the test is based on whether or not the proposed development or alteration will impact on the looks of an area or on the living standards of those who may be affected by it.

This is easy to understand when we are considering the building of a new property or the major extension of an existing property, but is perhaps less obvious when adding a fireplace or chimney to an existing property. Despite this, the requirement for planning permission may still apply. In some areas, such as conservation areas or national parks, or for some particular properties such as listed buildings, there are specific legislation and planning requirements and these are considered in the section below on 'Specific Legislation', but even in the average street on the normal semi-detached house there could still be a requirement.

Who is Responsible for Complying with the Regulations?

It is the responsibility of the property owner to ensure that if planning consent is required it is applied for and obtained. The contractor should ensure that the consents are in place before the work is started, but they will not be held responsible if the work is completed and then has to be changed or removed because planning consent has not been obtained.

There is also the problem of ensuring that the work carried out by the contractor complies with the planning consent obtained: for example, that the chimney does not end up taller than the consent allows; or that the finishing materials are of the agreed types; or that the location of the flue is in accordance with the approved plans. In all these cases, the contractor can be held responsible for any adjustments or changes if the contract between you and them has been correctly agreed, but despite this it is still the property owner who is held to be in breach of the Planning Regulations.

Who is Responsible for Policing the Regulations?

The local authority planning department is the organization that advises on and issues planning consent, and is also the organization responsible for ensuring compliance.

When do the Regulations Apply?

The actual need to gain planning permission will most likely pivot around whether or not the new chimney can be seen from the road or will interrupt someone else's view. The aspects of construction method are less likely to be considered at this stage as these will be dealt with under Building Regulations and Building Control, but the finishing materials or final appearance may be important.

BUILDING REGULATIONS

General

Building Regulations apply to every aspect of construction and should be followed whenever a new house, extension or conversion is carried out. In the past, it has been the law that the Regulations in force at the time of the construction of the property should be applied when refurbishment works are carried out to an existing property. However, in England and Wales during the year 2000 a new set of Building Regulations were announced and on 1 April 2002 they came into force. These introduced the principle that in the case of certain controlled works any refurbishment of existing facilities must be made to comply with the Building Regulations in force at the time the new work is carried out. As a result of these works being labelled 'controlled works', a number of additional regulations were applied. One of the most important stipulates that before these works are carried out there must be building control permission in place.

The types of work included under this new principle include such things as insulation, heating controls, replacement windows and doors, as well as the provision of a hearth, fire surround, new chimney and new chimney lining. The installation of any open fire, stove or chimney must therefore now comply with Building Regulations J1 to J4 and the refurbishment or recommissioning of an existing flue must comply with Regulations J2 to J4.

The Building Regulations are simple paragraphs that do not give any information on how to conform to them, so to help with compliance the Government has issued *Approved Document J: Guidance and Supplementary Information on the UK Implementation of European Standards for Chimneys and Flues*, which provides a lot of information on how to comply with the Regulations. This information is contained in a series of clauses known as 'deemed to satisfy clauses', and if the work is carried out in accordance with these it is 'deemed to satisfy' the Building Regulations J1 to J6.

Contained here are some extracts of non-technical clauses from *Approved Document J* to give an indication of the type of things they include. The extracts chosen are those that discuss the details of controlled works and building control permission; the technical issues involved will be considered when we go through the technicalities of various works later on in the book.

Who is Responsible for Complying with the Regulations?

Again, it is the responsibility of the property owner to ensure that if works are 'controlled works' the relevant permission is in place prior to the work going ahead. Even if the work is not controlled works, it is still the owner's responsibility to ensure that the work carried out is in compliance with the relevant Regulations.

It may seem a little unfair that if you employ a professional to carry out work for you and it proves to be controlled works and you do not have permission, it is you who will be breaking the law and liable for a fine, but that is indeed the case.

Who is Responsible for Policing the Regulations?

In the case of new build properties there are a number of possibilities as to who will police building control issues. The local authority is the first port of call, but if you employ a builder who is a member of the NHBC (National House-Building Council), then the NHBC will be responsible. Moreover, if the builder wishes to, he can contract out the building control policing to other organizations such as Zurich Insurance and so on. The law states that companies can register as independent building inspectors, and as a result can sign off work carried out under the scheme.

In the case of refurbishment it is more likely that the local authority will be the building control authority.

When do the Regulations Apply?

As shown above, the Building Regulations apply to almost all installations of hearths, fireplaces

and chimney linings. Accordingly, for the purpose of this book it is best to consider that they apply in all cases. However, if you make an approach to

the Building Control Department you will be told whether or not, the Department needs to be involved.

Extracts from Building Regulations *Approved Document J 2000, 2002 Edition (Combustion Appliance and Fuel Storage Systems) (ADJ)*

Repair of Flues

It is important to the health and safety of building occupants that renovations, refurbishments or repairs to the flue liners should result in flues that comply with the requirements of J2 to J4. The test procedures referred to in paragraph 1.53 and in Appendix E can be used to check this.

Flues are controlled services as defined in Regulation 2 of the Building Regulations, that is to say they are services in relation to which Part J of Schedule 1 imposes requirements. If renovation, refurbishment or repair amounts to or involves the provision of a new or replacement flue liner it is 'building work' within the meaning of Regulation 3 of the Building Regulations. 'Building work' must not be undertaken without prior notification to the local authority. Examples of work that would need to be notified include:

(a) Relining work comprising the creation of new flue walls by the insertion of new linings such as rigid or flexible prefabricated components.
(b) A cast in situ liner that significantly alters the flue's internal dimensions.

Reuse of Existing Flues

1.36 Where it is proposed to bring a flue in an existing chimney back into use or to reuse a flue with a different type or rating of appliance, the flue and the chimney should be checked and, if necessary, altered to ensure that they satisfy the requirements for the proposed use. A way of checking before and/or after remedial work would be to test the flue using the procedures in Appendix E.

1.37 A way of refurbishing defective flues would be to line them using the materials and components described in Sections 2, 3 and 4 dependent upon the type of combustion appliance proposed. Before relining flues, they should be swept to remove deposits.

1.38 A flue may also need to be lined to reduce the flue area to suit the intended appliance. Oversize flues can be unsafe.

1.39 If a chimney has been relined in the past using a metal lining system and the appliance is being replaced, the metal liner should also be replaced unless the metal liner can be proven to be recently installed and can be seen to be in good condition.

Condition of Combustion Installations at Completion

1.53 Responsibility for achieving compliance with the requirements of Part J rests with the person doing the work. That 'person' may be, e.g. a specialist firm directly engaged by a private client or it may be a developer or main contractor who has carried out work subject to Part J or engaged a subcontractor to carry it out. In order to document the steps taken to achieve compliance with the requirements, a report should be drawn up showing that materials and components appropriate to the intended application have been used and that flues have passed appropriate tests. A suggested checklist for such a report is given at Appendix A and guidance on testing is given at Appendix E. Other forms of report may be acceptable. Specialist firms should provide the report to the client, developer or main contractor who may be asked for documentation by the building control body.

1.56 Where a hearth, fireplace, flue or chimney is provided or extended (including cases where a flue is provided as part of the refurbishment work) information essential to the correct application and use of these facilities should be permanently posted in the building. A way of meeting this requirement would be to provide a notice plate as shown in diagram 1.9.

Because it was recognized that this new regulatory principle would cause an excessive workload for the local authority building control departments, a secondary piece of legislation was introduced in the form of Statutory Instrument 2002/440! This is a very important piece of legislation which allows members of certain 'competent persons schemes' to self-certify their work without having to apply to building control departments first. Below is a list of the relevant competent persons schemes currently in place and the type of works they are able to self-certify.

CORGI

This scheme has been in place for many years and formed the basis of the self-certification scheme. For a long time it has been recognized that the installation of gas pipes and gas appliances is a specialist field and that this work must only be carried out by a competent person, that is, an installer who is registered with the Council for the Registered Gas Installers (CORGI). CORGI registration covers a wide range of installation types and practices, and it should be noted that not all registered engineers are qualified and registered for all the various categories of work. Accordingly, it is important that you ensure that the engineer who carries out any gas work in your property is registered for that type of work, for example an engineer who is registered to install gas fires and cookers may not be registered to install or service a gas boiler.

If the individual carrying out the work has the relevant registration, then a CORGI-registered company can install any gas appliance, gas fire, boiler or cooker, along with any flexible stainless steel lining that may be required without them having to gain building control permission, but they must issue you with a report of the installation and the relevant tests and results so that you can prove the installer was a competent person and that they signed off the work as fully complying with section J of the Building Regulations. A copy of the report and a diagram of a sample notice plate for when a fireplace, hearth or flue liner has been installed are provided overleaf.

OFTEC

This is the competent persons scheme for oil-fired appliances and so, as with CORGI, a company registered with OFTEC (Oil Firing Technical Association) can install all types of oil-fired heating appliances and any flexible stainless steel liners that are required to make them work.

The main difference in this case is that any individual is allowed to carry out the work, even if they are not registered, provided that they gain the required building control approval prior to the installation starting. This flexibility is in place to allow the installation of these appliances by individuals who wish to make the installation in their own property, or for the benefit of companies that specialize in, say, gas installations and only carry out a few oil installations per year, thus making registration a less attractive proposition.

It is still recommended that any oil appliance is installed by an OFTEC-registered installer; indeed, some appliance manufacturers make it a condition of the appliance warranty.

HETAS

This is the competent persons scheme for solid fuel appliances and chimneys. This scheme operates on the same basis as OFTEC, in that individuals can install a solid fuel appliance if building control approval is granted prior to the installation taking place. The main difference in this case is the inclusion of chimney liners and systems in addition to the flexible stainless steel linings. If a HETAS member is registered for the installation of other types of systems, such as prefabricated chimney systems or sectional and cast in situ lining systems, they can also carry out this work without prior building control permission. **It must be pointed out that if the construction or refurbishment of a chimney requires foundation work to be carried out then building control permission must always be obtained prior to work starting irrespective of the person doing the works.**

HETAS also registers the members of a trade association known as NACE (National Association of Chimney Engineers) and supervises the Association's quality-control scheme. As a result, members of this scheme are also able to carry out the types of chimney work for which they are registered as competent persons, so long as the finished work is for a solid fuel appliance being installed at the same time.

CHECKLIST

Hearths, Fireplaces, flues and chimneys

This checklist can help you to ensure hearths, fireplaces, flues and chimneys are satisfactory. If you have been directly engaged, copies should also be offered to the client and to the Building Control Body to show what you have done to comply with the requirements of Part J. If you are a sub-contractor, a copy should be offered to the main contractor.

1.	Building address, where work has been carried out	
	Property address and post code	
2.	Identification of hearth, fireplace, chimney or flue.	Multifuel stove fitted in the fireplace recess in the lounge
3.	Firing capability: solid fuel/gas/oil/all.	All solid fuels
4.	Intended type of appliance. State type or make. If open fire give finished fireplace opening dimensions.	Multi fuel closed appliance or high efficiency wood burning stove Harmoney 5 by Euroheat
5.	Ventilation provisions for the appliance: State type and area of permanently open air vents	none visible on inspection, not required under ADJ table 2.1
6.	Chimney or flue construction	
a)	State the type or make and whether new or existing	New masonry chimney with Isokern Liners liners
b)	Internal flue size (and equivalent height, where calculated - natural draught gas appliances only)	150mm int Ø
c)	If clay or concrete flue liners used confirm they are correctly jointed with socket end uppermost and state jointing materials used.	The liners are installed with the socket upermost and the joints have been made with Isokern jointing compound. Lecca has been used as an insulating back fill
d)	If an existing chimney has been refurbished with a new liner, type or make of liner fitted.	
e)	Details of flue outlet terminal and diagram reference.	
	Outlet Detail:	louvre pot with 200mm inner diameter at the top, bird guard fitted.
	Complies with:	ADJ diagram 2.1 B
f)	Number and angle of bends.	2 bends @ 45°
g)	Provision for cleaning and recommended frequency.	From the base of the flue. To be swept twice a year when burning other than smokeless fuels
7.	Hearth. Form of construction. New or existing?	The constructional hearth is formed by the solid floor screed and the decorative hearth is at least 300mm in front of fireplace opening
8.	Inspection and testing after completion Tests carried out by: Mr M I Waumsley Tests (Appx E in AD J 2002 ed) and results	Comments The lack of gather and size of flue limits this flue to use with a closed appliance only
Flue Inspection	visual	CCTV survey using Wohler vis 2000 system, all OK
	sweeping	Chimney swept through the appliance
	coring ball	125mm core ball passed down flue.
	smoke	N2 pressure test to BSEN 1443 2002 passed with 23% of allowable leakage
	Appliance (where included) spillage	Smoke draw test (ADJ smoke test two) all smoke entered flue

I/We the undersigned confirm that the above details are correct. In my opinion, these works comply with the relevant requirements to the Building Regulations.

Print name and title	Mr M I Waumsley	Profession	CONSULTANT
Capacity	Consultant	Tel no	01234 567890
Address	The house, The street, The town	Postcode	DE** ****
Signed		Date	today
Registered membership of... (e.g. CORGI, OFTEC, HETAS, NACE, NACS)	**HETAS registered consultant and inspector**		

Fig 47   A sample of the recommended report format in Approved Document J, *Appendix A, completed for a solid fuel stove installation.*

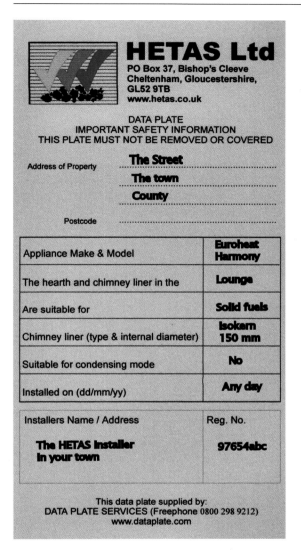

HETAS Ltd
PO Box 37, Bishop's Cleeve
Cheltenham, Gloucestershire,
GL52 9TB
www.hetas.co.uk

DATA PLATE
IMPORTANT SAFETY INFORMATION
THIS PLATE MUST NOT BE REMOVED OR COVERED

Address of Property	**The Street**
	The town
	County
Postcode	

Appliance Make & Model	**Euroheat Harmony**
The hearth and chimney liner in the	**Lounge**
Are suitable for	**Solid fuels**
Chimney liner (type & internal diameter)	**Isokern 150 mm**
Suitable for condensing mode	**No**
Installed on (dd/mm/yy)	**Any day**

Installers Name / Address	Reg. No.
The HETAS Installer in your town	**97654abc**

This data plate supplied by:
DATA PLATE SERVICES (Freephone 0800 298 9212)
www.dataplate.com

Fig 48   A sample of a notice plate completed for the installation documented in the report.

The above schemes are there for the protection of the consumer to ensure that an installation carried out by those companies registered with the respective scheme is both safe and complies with the relevant regulations. **The schemes provide both quality-control regimes and complaints procedures, and so I would strongly recommend that installations are carried out by people registered with the**

relevant body and that this book is used as a guide by readers to ensure they get the best possible installation for their requirements.

BRITISH STANDARDS (AND EUROPEAN NORMATIVE STANDARDS)

General

These are not regulations, nor are they compulsory unless they are deemed necessary by either the Building or Planning Regulations. An example of this can be seen below where ADJ requires compliance with a British Standard. BS stands for a British Standard; BS EN stands for a European Normative Standard that has been adopted by Britain; and prEN stands for a proposed European Standard.

Construction of Masonry Chimneys

1.27 New chimneys should be constructed with flue liners and masonry suitable for the intended application. Ways of meeting the requirement would be to use bricks, medium weight concrete blocks or stone (with wall thicknesses as given in Sections 2, 3 or 4 according to the intended fuel) with suitable mortar joints for the masonry and suitably supported and caulked liners. Liners suitable for solid fuel appliances (and generally suitable for other fuels) could be:

a) liners whose performance is at least equal to that corresponding to the designation T450 N2 S D 3, as described in BS EN 1443:1999, such as:

   i) clay flue liners with rebates or sockets for jointing meeting the requirements for Class A1 N2 or Class A1 N1 as described in BS EN 1457:1999; or

   ii) concrete flue liners independently certified as meeting the requirements for the classification Type A1, Type A2, Type B1 or Type B2 as described in prEN 1857(e18) January 2001; or

   iii) other products that are independently certified as meeting the criteria in a).

b) imperforate clay pipes with sockets for jointing as described in BS 65: 1991 (1997).

Who is Responsible for Complying with the Standards?

In most cases a BS or BS EN is a code of practice or an advisory document, unless a contract or quotation specifically refers to them. In the event of an installation failing, the installation may be surveyed and the surveyor may report that the installation does not comply with good practice because it does not meet the requirements of a relevant British Standard. However, this would be purely a civil matter and the surveyor is only indicating that had the installation complied with the relevant BS it would have been more likely to have worked.

Who is Responsible for Policing the Standards?

In the case of installation codes of practice, as opposed to product Kite markings, there is no policing body. The only exceptions are the quality scheme standards such as the BS 5750 and ISO 9000 series. In these cases, if an installation company claims to have a quality standard that requires them to apply British Standards, and they fail to do so, then the company issuing the quality certificate can be asked to investigate this 'quality failing'.

When do the Standards Apply?

A relevant British, or European, Standard can be referred to in any contract document or quotation for work and so can be made to apply. If the Standard is referred to in a particular Building or Planning Regulation, it must be followed whenever that Regulation applies.

SPECIFIC LEGISLATION

General

A full explanation of this type of legislation is beyond the scope of this book and so only a brief explanation has been provided. For further information, please contact the relevant governing bodies.

Gas Safety in Use Regulations

Any installation of a gas service or appliance must comply fully with these Regulations and any amendments made to them. These are policed by the Health and Safety Executive and include wide-ranging use of British Standards and specific regulations. These are rigorously enforced and failures to comply will likely result in a prosecution of the person responsible.

Listed Buildings

A number of buildings in the UK are listed buildings. A building may have been listed because of its special architectural interest, its unique or unusual construction methods, or the fact that it is of historic significance. Properties are normally listed in one or other of the following categories, according their importance, and the category will dictate the type of works or alterations that can be carried out. Grade II listed, Grade II* listed and Grade I listed buildings will all have restrictions over and above the normal Planning Regulations, in some instances down to the colour of paint or whether the owner is allowed to change the internal décor. It is therefore important that if you are the owner of such a building you obtain listed building consent prior to any work being started. There is a register of listed buildings if you are unsure whether or not the building is listed, and there is also an association for owners of such buildings, the Listed Property Owners Club, that can provide advice and support should you need to make an application for changes. Their contact details are given in 'Useful Addresses' below.

Conservation Areas

These include areas such as the Peak District National Park and World Heritage sites and so on. Each of these areas will have its own planning committee or planning board, which will have established its own criteria for the acceptance of various styles or materials based on the prevailing styles in its area of jurisdiction.

These criteria can often be very difficult to meet and it is essential that any changes are cleared prior to starting any work. The planning boards dealing with conservation areas have wide-ranging powers, and failure to obtain, or comply with, planning approval can result in the enforced removal of any additions you have made to a property.

CHAPTER 6

Choice of Chimney

The choice of chimney construction and materials will depend on many things, not only the requirements of the appliance (which have already been discussed and will be covered in more detail later), but also the circumstances in which the chimney is being constructed. I have categorized these varying circumstances below, and under each category I discuss the type of chimney that is suitable and the correct installation procedures involved.

IN NEW HOUSES

In new houses it is still common practice to use the traditional materials and construction methods for a chimney, particularly where this chimney is intended for solid fuel use. This is often due to the lower cost as all the required trades are already on site and the materials also are cheap. If this type of construction is designed and specified well and is built with care, then it is a good chimney that will be suitable for any fuel type.

In some specialist constructions the purpose-designed factory-made block systems are used. These should provide for an easier and quicker build with a more insulated and often more resilient chimney. These systems are designed to be built straight, but they all provide the facility to include offsets if necessary; however, these should be avoided wherever possible.

The chimney types described above should last at least twenty-five years and in many cases will have a life equal to that of the dwelling design (approximately sixty years).

In some properties, such as barn conversions or oak timber-framed houses, a stainless steel or similar prefabricated chimney is often used. This is because they take less room, are easier to install and are often considered for their aesthetic appeal in some settings. In the case of barn conversions, especially in conservation areas, this type of chimney is preferred by planners because when painted black they are less obtrusive and have less effect on the shape of the building.

These prefabricated flues have a life of between ten and twenty years and should be fully inspected after ten years of use and any damaged sections replaced.

ADDED TO EXISTING HOMES

The addition of a new chimney to an existing property is a very different proposition to including one in a new home. In this case, using a prefabricated chimney tends to be preferable to constructing a chimney from scratch. The reason for this is that it can erected and connected without the need for any changes to the foundations. Prefabricated flues are relatively light and all their weight can be supported on brackets as they pass through ceilings and roofs. An additional advantage is that they are relatively small and because they are insulated they can be put through combustible floors with relatively small holes and limited trimming of joists.

The factory-made block systems are produced using lightweight insulating concrete, often including aggregates made from pumice, a volcanic rock with a very high temperature resistance and insulation value. Both of the above flues are smaller in dimension than a brick-constructed chimney with the same flue liner in it and both will provide a more insulated chimney, as well as be quicker and easier to install. A heating installer will often choose a prefabricated chimney system as it does not require a brick layer or the use of mortar and so on.

The construction of a traditional masonry chimney will involve costs for changes to the foundations and the construction of a brick layer, as well as the fees of the appliance installer. Indeed, labour charges will be quite high as it takes a considerable amount of time to add a masonry chimney to an existing property.

IN SPECIAL CIRCUMSTANCES

There are occasions when the type of building will dictate the type of flue that can be used, for example when a chimney is installed into a mobile home or a property that has foundations based on a piling system. Even in timber-framed buildings it is better to use a factory-made block system, as such a system incorporates provision for reinforcements that will enable a greater unsupported length of flue. These systems also include a precast appliance chamber system which will facilitate the safe installation of any type of appliance.

To give a better understanding of which type may be the best in your own particular circumstances, the next sections look at the methods of construction and at some issues raised by the Building Regulations regarding the relevant construction methods.

PRELIMINARY DECISIONS AND ACTIONS

Flue Size

If an open fire is being considered, then the final fireplace opening must be decided at this early stage to ensure that the flue being constructed will not be too small, or, equally as bad, too large for the fire to be installed.

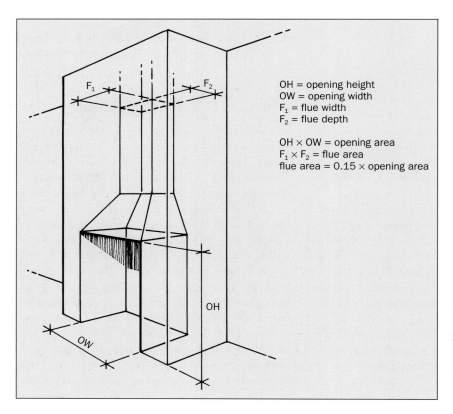

OH = opening height
OW = opening width
F_1 = flue width
F_2 = flue depth

OH × OW = opening area
F_1 × F_2 = flue area
flue area = 0.15 × opening area

Fig 49   The area of the flue should be 15 per cent of the fireplace opening area according to Approved Document J *for England and Wales.*

ADJ puts flues serving open fires into two different categories: those serving fireplaces up to 500 × 550mm (19½ × 21½in); and those serving fireplaces larger than this. First, we will look at flue sizes for those serving fireplaces up to 500 × 550mm (19½ × 21½in).

If the appliance to be installed is a decorative fuel effect (DFE) gas fire, then the minimum flue size must be 175mm (6⅞in) in dimension whether it is round or square.

If the appliance to be installed is a solid fuel open fire, then the minimum flue size is 200mm (8in) diameter, or a rectangular/square flue having the same cross-sectional area and a minimum dimension of 175mm (6⅞in).

The flue size for the solid fuel fire is larger as the sooty deposits will reduce the area as they build up – if the flue size was only 175mm (6⅞in) diameter it would require sweeping after almost every use. It can be seen from the above that a fireplace recess designed to suit a solid fuel appliance with a flue of 200mm (8in) internal diameter can also be used for a DFE, as the 175mm (6⅞in) dimension given is only a minimum requirement.

For a fireplace with dimensions greater than 500 × 550mm (19½ × 21½in) the requirements under the Building Regulations are the same for a DFE as they are for a solid fuel open fire, but they differ in England and Wales to those in Scotland. In England and Wales a very simplistic attitude is taken, which is that the area of the flue should equal 15 per cent of the area of the final fireplace opening. As this takes no account of the height of the flue it is a very inaccurate and potentially dangerous method of sizing a flue for an open fire.

In Scotland the Building Regulations provide a graph for the sizing of flues for open fires greater than 500 × 550mm (19½ × 21½in). This takes into

---

Differences in Flue Size When Using the 15 Per Cent Rule or the Sizing Graph

The table below illustrates the differences that can exist between using the 15 per cent rule and one of the flue sizing graphs. As an example, to size a flue for a fireplace of 800 × 900mm (31 × 35in), the two methods will result in very different flue sizes:

Property Type	15 Per Cent Rule			From Graph		
	Sectional Area (m^2)	Surface Area (m^2)	Volume (m^3)	Sectional Area (m^2)	Surface Area (m^2)	Volume (m^3)
Bungalow	0.105	5.25	0.47	0.065	0.424	0.029
Two-storey house	0.105	9.33	0.84	0.05	6.28	0.4
Three-storey house	0.105	12.6	1.13	0.045	7.63	0.49

The graph reflects the effect of a flue's height on a chimney's capabilities, and the fact that it will have sufficient energy to draw the gases through the flue faster and so will require a smaller liner.

The effect the height of a flue on the surface area and volume of the flue can also be seen in the table above, and this will affect the heat loss from the flue. Using only the 15 per cent rule is likely to cause a significant oversize of the flue and a corresponding decrease in flue temperature, resulting in the flue gases getting so cool that they will begin to fall in the flue and cause the fire to smoke back.

In the England and Wales *Approved Document J* there is a clause that allows the designer to seek expert help for large flues and this can be done by using the graph from the Scottish Regulations or the one available from the Solid Fuel Association.

Minimum Flue Sizes for Solid Fuel Appliances	
Appliance Description	**Minimum Flue Size**
Closed appliance up to 20kW rated output that burns smokeless fuel	125mm (5in) internal diameter or a square flue having the same cross-sectional area
Closed appliance up to 30kW rated output burning any fuel	150mm (6in) internal diameter or a square flue having the same cross-sectional area
Closed appliance aove 30kW and up to 50kW rated output burning any fuel	175mm (7⅞in) internal diameter or a square flue having the same cross-sectional area

account the height of the flue and so will give a more accurate assessment of the flue size required. The Solid Fuel Association also provides a graph.

When considering a closed appliance such as a multi-fuel stove, then the appliance manufacturer's installation instructions must be studied and their recommendations considered alongside the Building Regulations.

ADJ and the Scottish Regulations agree on the sizing of flues for solid fuel appliances and the table above provides the information required to ensure that the minimum sizes are adhered to when choosing a flue size. This table indicates that any solid fuel appliance, including a wood-burning stove fitted to a flue of 125mm (5in) internal diameter, is restricted to burning only smokeless fuel in the UK. For this reason, I would recommend that a 125mm flue is only used where a smokeless-fuel-only appliance is fitted, or in a smoke-control area where only smokeless fuel can legally be burned.

Fireplace Recess Type

Having chosen our appliance, and hence the flue size, we now move on to deciding on the type of recess. If a fireplace opening of greater than 500 × 550mm (19½ × 21½in) has been chosen, the fireplace recess must be designed to take this, including the gather and the hearths; we will come back to this later. If a closed appliance or an open fire of less than 500 × 550mm (19½ × 21½in) has been chosen, we can consider a universal opening.

The illustrations opposite show a universal opening and how it can be adapted to suit either an open fire or a closed appliance. The width of the opening can be greater than the sizes shown, but the height must be within the tolerances indicated if it is going to be capable of being adapted to suit an open fire.

If a greater height is required to give a more imposing fireplace for a closed appliance then it becomes specific to a closed appliance and can no longer be considered a universal opening, because the components to convert it to an open fire will no longer fit into it correctly.

Many customers request an inglenook opening. However, this can lead to confusion in that while some do require an inglenook, others in fact just want a large fireplace opening.

Figure 52 shows an example of a true inglenook, which takes the form of a small annex to the main room, often with seats down either side, with a smaller fireplace either free-standing in the space or set into the back wall. This is easy to achieve with a stove, but requires more thought when considering an open fire.

There are a number of ways of providing an inglenook with an open fire. These include using a free-standing open fire such as the Dovré 2000 series, or similar, or providing a canopy over the grate to restrict the actual fireplace opening area to the canopy perimeter. The way to calculate the area under a canopy is given in Figure 55.

Alternatively, a brick chimney breast can be constructed within the inglenook or the fireplace can be recessed into the back wall of the inglenook. Both of these suggestions will reduce the size of flue liner

Fig 50   Universal builders opening set out for use with a free-standing stove.

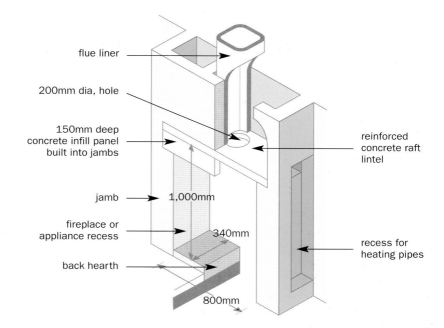

flue liner

200mm dia, hole

150mm deep concrete infill panel built into jambs

jamb → 1,000mm

fireplace or appliance recess

340mm

back hearth

800mm

reinforced concrete raft lintel

recess for heating pipes

Fig 51   A universal builder's opening adapted for an open fire using precast components.

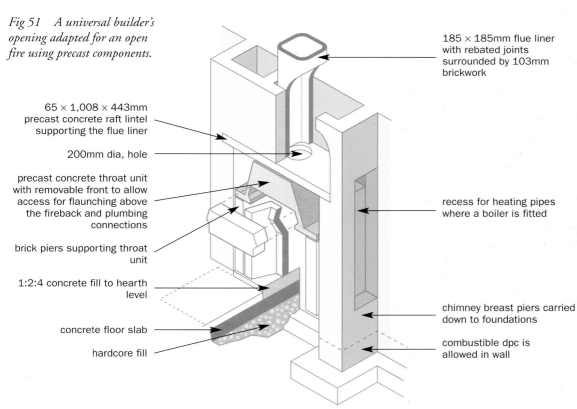

65 × 1,008 × 443mm precast concrete raft lintel supporting the flue liner

200mm dia, hole

precast concrete throat unit with removable front to allow access for flaunching above the fireback and plumbing connections

brick piers supporting throat unit

1:2:4 concrete fill to hearth level

concrete floor slab

hardcore fill

185 × 185mm flue liner with rebated joints surrounded by 103mm brickwork

recess for heating pipes where a boiler is fitted

chimney breast piers carried down to foundations

combustible dpc is allowed in wall

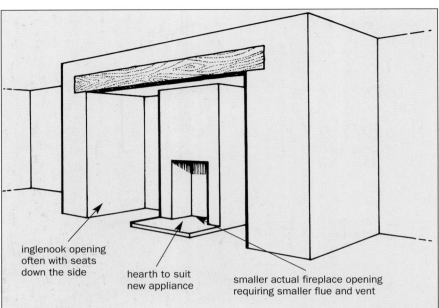

inglenook opening
often with seats
down the side

hearth to suit
new appliance

smaller actual fireplace opening
requiring smaller flue and vent

Fig 52   A true inglenook is a small room off the main room with a smaller fire set into it.

Fig 53   The Dovré 2000 series set in a large opening to give an inglenook-type fireplace.

Fig 54   A Catchsmoke canopy, manufactured by Enginuity Developments, fitted over a large grate will reduce the flue size and form an inglenook.

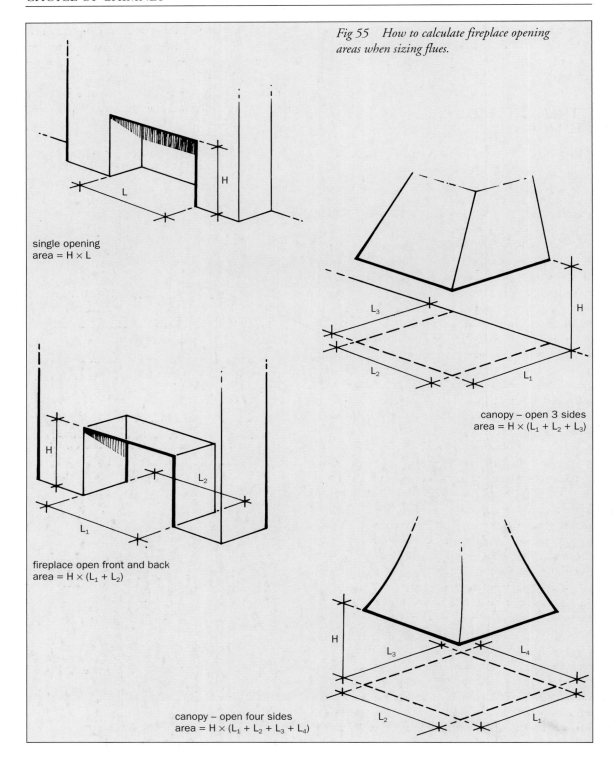

Fig 55 How to calculate fireplace opening areas when sizing flues.

single opening
area = H × L

fireplace open front and back
area = H × (L₁ + L₂)

canopy – open 3 sides
area = H × (L₁ + L₂ + L₃)

canopy – open four sides
area = H × (L₁ + L₂ + L₃ + L₄)

required and the amount of air taken out of the room up the chimney, as opposed to just having a large fireplace with a coal grate or log basket sitting within it.

A TRADITIONAL BRICK CHIMNEY

This is still the most common form of construction, with the cheaper clay liners being the most commonly used. This type of construction will provide greater flexibility with regard to the fireplace and flue size and so is often used where a large fireplace or inglenook is required.

There are few manufacturers who provide detailed installation instructions for their products in relation to the construction of the flue as a whole, and so most of the construction detail needs to be obtained from the relevant British Standards and the Building Regulations. This section is intended to provide the details you will need in order to be able to check on a design being offered by a professional and to help to identify any problems as the construction is taking place. **The construction of this type of chimney requires the specialist skills of a competent bricklayer and so this section is not intended as a set of installation instructions.**

Firstly, it is important to obtain a full set of calculations and drawings for the foundations of the chimney, based on the proposed height of the chimney, the internal size of the flue, the size of the fireplace and the materials to be used.

From the foundations, a fireplace recess has to be constructed and this can be done in two ways, as discussed above, either as a specialist recess designed specifically for the appliance or fireplace to be used, or as a universal fireplace recess. The first method has the advantage of being specific to your needs and meeting the full requirements of the flue from this point upwards to ensure correct design, while the other method provides some flexibility with regard to the final appliance installation.

In this case, we will consider the use of an open fire less than 500 × 550mm (19½ × 21½in) in a fireplace recess designed specifically for an open fire. The standard dimensions for the fireplace recess are 600mm wide × 350mm deep × 600mm high (24 × 14 ×

24in). The thicknesses of the walls forming this recess are dictated by the Building Regulations, but in general for a chimney that is suitable for all fuel types the wall thicknesses should be 200mm (8in).

In the base of the recess, and in front of it, there must be a constructional hearth. This must

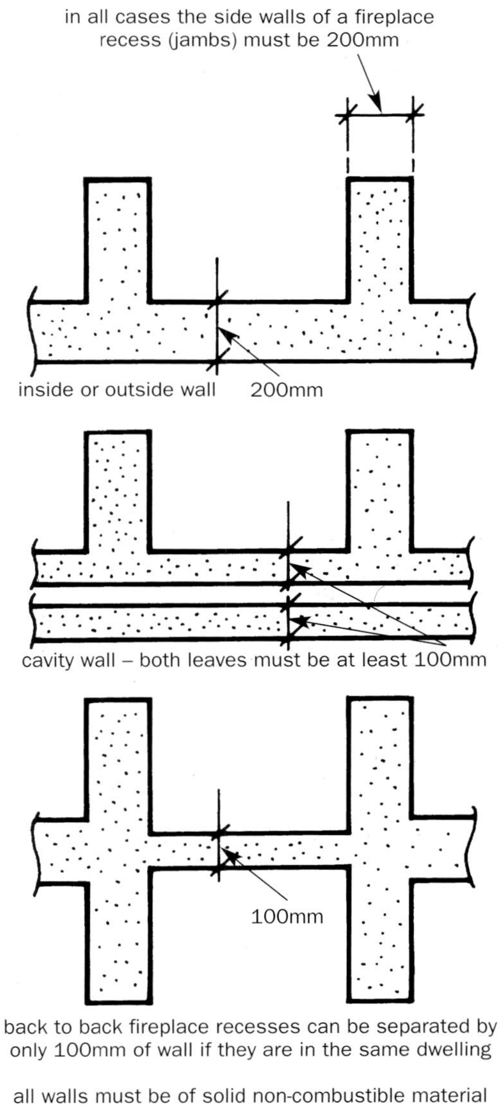

Fig 56   The wall thickness for a fireplace recess for solid fuel (also suitable for other fuels).

69

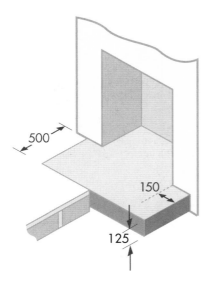

Fig 57   A constructional hearth in a fireplace recess.

completely fill the recess and extend in front of it by 500mm (19½in) and either side of the recess by at least 150mm (6in). This constructional hearth must be at least 125mm (5in) thick of solid non-combustible material and cannot have any combustible material below it unless there is an air gap of at least 50mm (2in) or 250mm (9¾in) of solid material.

At the top of this recess a precast concrete gather unit should be installed to provide both the gather formation and the throat-forming lintel. Some of these units have a removable throat-forming lintel to ease the installation of the fireback later; this is a good idea. The gather unit normally finishes in a large square top to enable an adaptor or starter block to be fitted so that a smooth change can be made from the standard gather to whatever liner size and shape has been chosen.

This gather unit is load-bearing and so the front of the gather unit should be placed level with the front

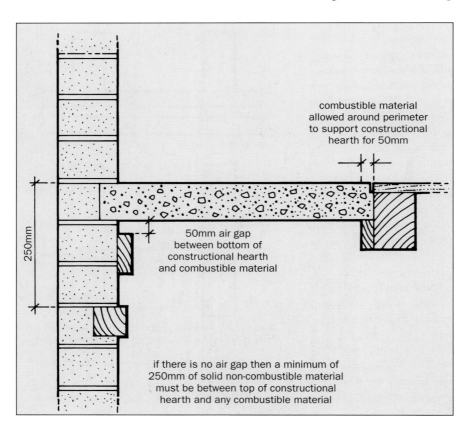

Fig 58 Combustible material under or supporting a constructional hearth.

70

Fig 59   A precast gather unit with a throat lintel and starter blocks.

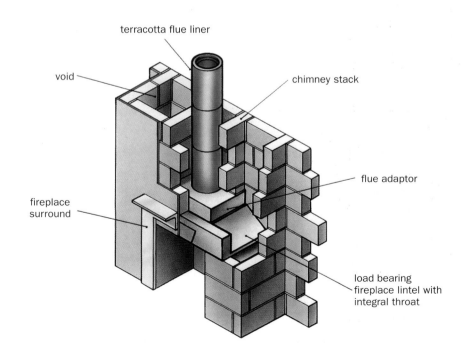

terracotta flue liner

void

chimney stack

fireplace surround

flue adaptor

load bearing fireplace lintel with integral throat

Fig 60   The construction of a chimney breast above a precast gather unit.

71

of the chimney breast in order for the throat-forming lintel to be located correctly. The top of the chosen adaptor or starter block will have an opening with a rebate to house the required flue liner. This will ensure a smooth change from the throat to the liner and that the flue liner is fully supported, the correct way up and sealed correctly onto the gather.

The brickwork to surround the liner should be constructed on the top of the gather as this will provide the correct support and size of chimney. To maintain the correct size of chimney breast in the room, the walls forming the sides of the chimney breast can then be continued upwards, forming a void.

The feature box contains an extract from *ADJ*. An explanation now follows of the main points of this extract.

All flue liners used in the UK should have a socket on one end and a spigot on the other. This is to enable the liners to be located securely on top of each other, to ensure they are correctly lined up and to enable a good seal to be made. The liners must be fitted with the socket uppermost to ensure that any acidic condensate formed in the flue, or any rain that enters the flue, is not allowed to leak out to attack the masonry surrounding the liner, which would cause damp and stained patches.

When specifying a liner the fuel type must be known, but if a liner suitable for solid fuel is specified this is in general suitable for all other combustion fuels. The only exception to this is where a condensing appliance is installed. The specification for this type of liner is T450 N2 S (or G) D 3. An explanation of what this all means may be

---

Extract from *Approved Document J* Regarding the Construction of Masonry Flues

1.27 New chimneys should be constructed with flue liners and masonry suitable for the intended application. Ways of meeting the requirement would be to use bricks, medium weight concrete blocks or stone (with wall thicknesses as given in Sections 2, 3 or 4 according to the intended fuel) with suitable mortar joints for the masonry and suitably supported and caulked liners. Liners suitable for solid fuel appliances (and generally suitable for other fuels) could be:

a) liners whose performance is at least equal to that corresponding to the designation T450 N2 S D 3, as described in BS EN 1443:1999, such as:

   i) clay flue liners with rebates or sockets for jointing meeting the requirements for Class A1 N2 or Class A1 N1 as described in BS EN 1457:1999; or

   ii) concrete flue liners independently certified as meeting the requirements for the classification Type A1, Type A2, Type B1 or Type B2 as described in prEN 1857(e18) January 2001; or

   iii) other products that are independently certified as meeting the criteria in a).

b) imperforate clay pipes with sockets for jointing as described in BS 65: 1991 (1997).

1.28 Liners should be installed in accordance with their manufacturer's instructions.

Appropriate components should be selected to form the flue without cutting and to keep joints to a minimum. Bends and offsets should only be formed with matching factory-made components. Liners need to be placed with the sockets or rebate ends uppermost to contain moisture and other condensates in the flue.

Joints should be sealed with fire cement, refractory mortar or installed in accordance with their manufacturer's instructions. Spaces between the lining and the surrounding masonry should not be filled with ordinary mortar. In the absence of liner manufacturer's instructions, the space could be filled with a weak insulating concrete such as mixtures of:

a) one part ordinary Portland cement (OPC) to twenty parts suitable lightweight expanded clay aggregate, minimally wetted; or
b) one part OPC to six parts vermiculite; or
c) one part OPC to ten parts perlite.

helpful when considering the type of liner to be installed:

- T450 – this is the nominal flue gas temperature in °C;
- N2 – because flues are constructed of jointed components there will always be a risk of leakage and this figure gives the test criteria for the pressure testing of a product; it stands for natural draught criteria 2;
- S or G – this indicates that the flue liner has been tested to ensure it is capable of containing a soot or chimney fire;
- D – this means that the liner is designed for dry or non-condensing appliances only;
- 3 – this number indicates the grade of acid resistance provided; the lowest level is 1 and the highest level is 3.

The joints between the liners must be made with an acid-resistant jointing compound. Some liner manufacturers market their own compound and, if so, that particular brand must be used. If round liners are being used then the liners should be rotated in the socket until the best fit is obtained. The best-fit issue is especially important with clay liners as these will shrink and change shape during the firing stage of manufacture. They will therefore not be perfectly straight and round, and if they are just laid upon each other there will be a series of ledges and uneven joints that will affect the resistances in the flue, possibly even resulting in weak spots. Always use the longest liners available in the given diameter you require. This will reduce the number of joints and so reduce the risk of leaks and improve the surface finish of the inside of the flue.

The joint must be full of jointing compound, but when the flue liners are placed together any excess compound squashed out of the joint must be removed and smoothed off to avoid ledges and protrusions. At the end of each working day the bricklayer should draw a core ball or a bag of sand up the flue to ensure that any excess compound is removed or smoothed against the side of the liner before it sets hard and becomes difficult to remove.

The space between the brickwork forming the chimney and the outside of the liner must be fully filled with an insulating cement of the correct type and mix after each liner has been placed and not left till the end, as the material will not adequately fill the space if it is poured down later. The type of insulating fill used around the liners will depend on the liner material and the manufacturer's instructions must be followed. In general, for clay liners the mixes will be either 10:1 perlite to OPC, or 6:1 vermiculite to OPC. In both cases, a small amount of water should be put into the mix to ensure that it sets and does not settle to leave an uninsulated space at the top of the flue. If concrete or pumice liners are used, then the recommended insulating material is an expanded clay bead called leca, which should be mixed in a 20:1 ratio with OPC and a little water.

The Building Regulations require the use of an insulating back fill for two reasons: to increase the heat retained in the flue; and to ensure that a liner is held firm within the brick structure so that when the flue is swept, which could be more than 120 times during its lifetime, the liner will not be moved, resulting in damaged joints and leaks.

The chimney liners must be contained in a brick or block structure and in the case of a solid fuel chimney the thickness must be at least 100mm (4in) of solid non-combustible material around the flue liner and between two flues in the chimney. If the flue passes through any separate fire compartment, such as a flat above or a garage, then the walls at this point should be 200mm (8in) thick. Where a chimney passes through a floor or ceiling, then the floor joists or roof rafters must be kept 40mm (1½in) away from the chimney structure, or the chimney must have a wall thickness of at least 200mm. Any decorative timber can, however, be placed directly onto the chimney structure as long as the 100mm of solid non-combustible material is present.

As the chimney passes through the roof there are a number of issues to consider, such as weather protection, the height above the roof, the location of dormer windows, skylights or vents, and the location of the top of the chimney in relation to the ridge or other buildings. Again, the Building Regulations give us some assistance in this, as do the British Standards.

Figure 63 shows the correct way to flash the chimney into the roof and the location of

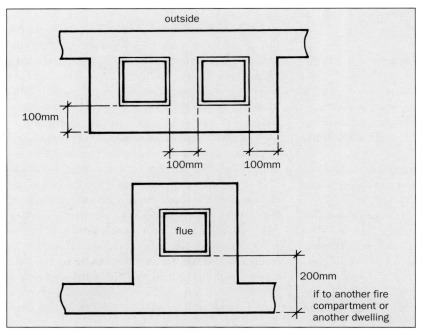

Fig 61   The thickness of the walls around a flue liner in a masonry chimney.

outside

100mm

100mm   100mm

flue

200mm

if to another fire compartment or another dwelling

Fig 62   The separation of combustible material from a flue in a chimney.

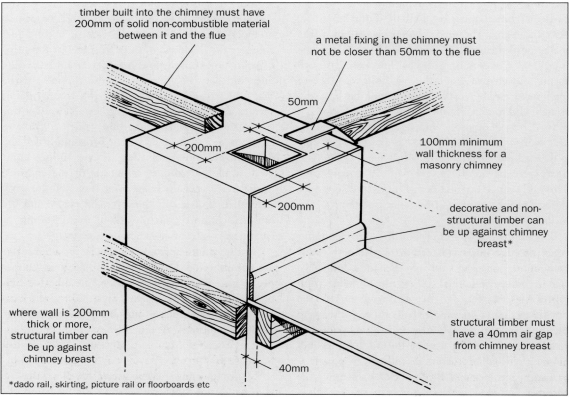

timber built into the chimney must have 200mm of solid non-combustible material between it and the flue

a metal fixing in the chimney must not be closer than 50mm to the flue

50mm

200mm

100mm minimum wall thickness for a masonry chimney

200mm

decorative and non-structural timber can be up against chimney breast*

where wall is 200mm thick or more, structural timber can be up against chimney breast

structural timber must have a 40mm air gap from chimney breast

40mm

*dado rail, skirting, picture rail or floorboards etc

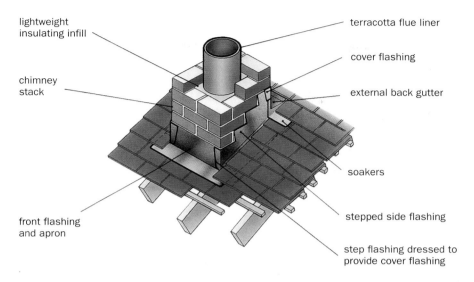

lightweight insulating infill

chimney stack

front flashing and apron

terracotta flue liner

cover flashing

external back gutter

soakers

stepped side flashing

step flashing dressed to provide cover flashing

Fig 63   Styles of flashing on chimney stacks.

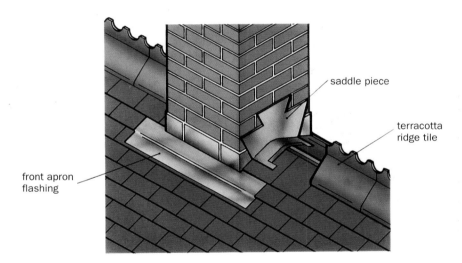

saddle piece

terracotta ridge tile

front apron flashing

damp-proofing and weather trays. Further details on this can be obtained from the Building Research Establishment (BRE), whose details can be found in 'Useful Addresses'.

Before considering where we will terminate the chimney we must consider the maximum height we are allowed to build up our chimney stack and the minimum height of flue we should have. There is a maximum unsupported height of chimney allowed to ensure the structure remains stable in all condi-

tions and this is dictated by the code of basic data for the design of buildings (CP3). The section of unsupported roof must be as short as is allowable to avoid any cooling effects on the exposed section as much as possible, but occasionally a chimney has to be tall to avoid areas of turbulent air and high-pressure regions caused by the roof or other tall structures. Where this is the case, the structure must not be taller than four and a half times the smallest horizontal dimension at the last point of support.

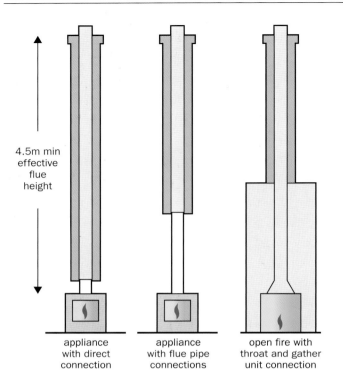

Fig 64   The minimum effective flue height for a multi-fuel chimney is 4.5m (15ft).

4.5m min effective flue height

appliance with direct connection

appliance with flue pipe connections

open fire with throat and gather unit connection

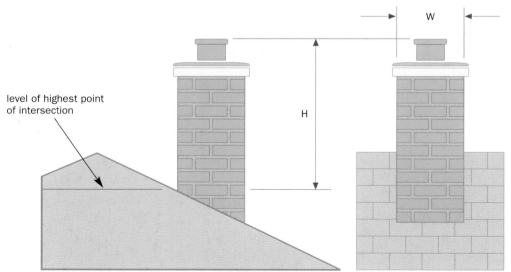

W

level of highest point of intersection

H

H should not exceed 4.5W. Note: W is the least horizontal dimension of the highest point of intersection with roof surface, gutter etc., and H is measured to top of any chimney pot or terminal

Fig 65   The maximum unsupported height of a chimney stack above the roof.

The last point of support will be the exit from the roof surface. If it is a pitched roof it is the higher point and so it is from this point that we measure vertically. In a standard chimney stack with one flue and a 225mm (9in) internal diameter flue liner, the horizontal dimensions of the chimney are likely to be 550mm (21½in) and so the maximum allowed stack height would be 2,200mm (86in), measured to the top of the chimney pot or terminal.

Now we have established the maximum height we need to know the minimum we are able to get away

with and this is again dictated in the Building Regulation, and BS 6461 'Installation of Chimneys and Flues for Domestic Appliances Burning Solid Fuel' gives further information to help when taking into account other factors such as topography, trees and so on.

The first thing to identify is where the chimney will come through the roof and what is around this point. The illustrations here show: the requirements of the Building Regulations for solid fuel chimneys passing through a slated or tiled roof; the re-

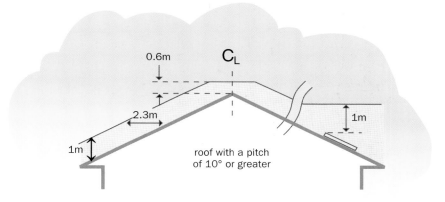

Fig 66 The minimum terminal locations of chimneys for solid fuel passing through a standard roof surface.

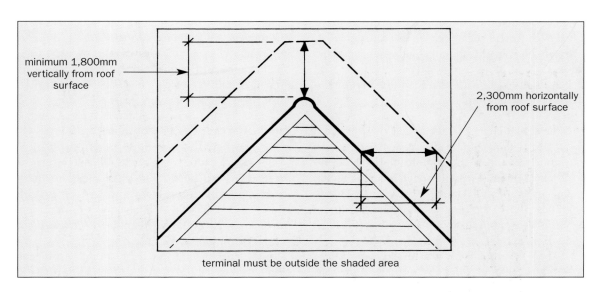

Fig 67 The minimum terminal locations of chimneys for solid fuel passing through an easily ignitable roof surface such as a thatched roof.

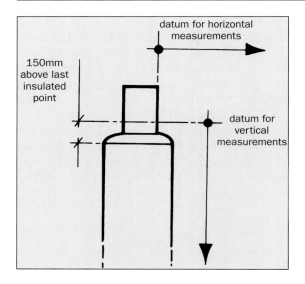

Fig 68   The datum points for horizontal and vertical measurements on a chimney.

quirements for chimneys passing through easily ignitable roof surfaces such as thatch and shingles; and the point in the chimney and its pot that measurements are to be taken to. The horizontal measurements are to be taken to the outside face of any chimney pot and the vertical measurements are to be taken to the top of the brickwork in Scotland and to 150mm (6in) up the chimney pot in England and Wales.

For a roof of normal construction such as tiles or slates the following applies:

- Where a chimney passes through the ridge of a roof or within 600mm (23in) of the ridge, the chimney should extend 600mm above the ridge.
- Where a chimney passes through a sloping roof at any other point it must be at least 1,000mm (39in) above the roof surface and 2,300mm (90in) away from the roof surface when measured horizontally.
- If there is a roof light, vent or dormer window within 2,300mm (90in) of the chimney, then the chimney terminal should be 1,000mm (39in) above the highest opening in the roof light, vent or dormer window.

- If the chimney is within 2,300mm (90in) of an adjacent or adjoining building, whether or not it is beyond the boundary, then the flue must extend 600mm (23in) above this building.

If the roof surface is of an easily ignitable material then the chimney must terminate further away from the roof surface, and so the notes above apply but with the following additional information:

- Where a chimney passes through the roof surface within 2,300mm (90in) of the ridge of the roof, then it must extend at least 1,800mm (70in) above the roof surface and 600mm (23in) above the ridge (note these are not alternatives – both must be satisfied).
- Where a chimney passes through the roof surface more than 2,300mm (90in) from the ridge of the roof, then it must extend at least 1,800mm (70in) above the roof surface and be at least 2,300mm horizontally away from the roof surface.

In the case of chimneys limited strictly to gas and oil-fired appliances these dimensions may vary and balanced flues may also be used.

We now reach the point where we have to consider what we are going to use to terminate both the chimney and the flue. There are two recognized ways to terminate the chimney structure: one is to use a proprietary chimney cap that extends over the edge of the chimney structure and includes a groove to prevent rain from returning under to meet the brickwork below; the alternative is to provide a drip course by corbelling the top two or three courses, using a damp-proof course of slate or similar, or using engineering brick for these top three courses.

Choosing a chimney pot can be difficult, especially for solid fuel appliances. Some gas appliances require specific types of terminals and these are dictated by either the appliance manufacturer's installation instructions or the Gas Safety in Use Regulations via various British Standards.

In the case of solid fuel there are many different chimney pots and terminals, so deciding on the correct one can be difficult, but in the first instance we should consider any information provided in the British Standards. BS 6461 states that for an open fire

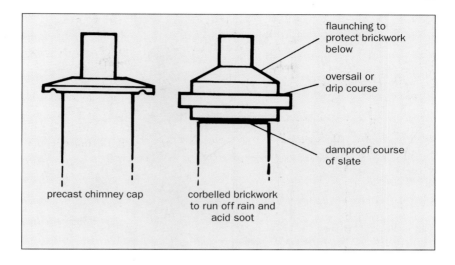

Fig 69   Chimney terminal types for masonry chimneys.

the outlet diameter of the flue terminal must be at least 200mm (8in) diameter or 185mm (7in) square, and in any case should be of the same dimensions as the flue. This at least tells us that if we have a round liner we should use a round chimney pot, and if we have a square liner we should have a square chimney pot. The sides of the pot should be parallel to ensure that the gases are not compressed and made to speed up at this point or they could cause pressure further down the flue. A minimum recommended chimney pot height is 150mm (6in); this is to position the terminal of the flue out of the turbulence caused around the brick stack on a windy day.

If a pot is very tall it will get cold as there is no insulation within it, which could cause rapid tarring or sooting up of the pot. A recommended maximum extension above the chimney is 450mm (17½in). This is sometimes extended where a specialist anti-downdraught pot is used to avoid turbulent air when a chimney terminates in a downdraught zone.

We can see that the ideal pot is one that is the same size and shape as the chimney liner and extends at least 150mm (6in) above the top of the chimney, but should not extend more than 450mm (17½in) unless there is a need to do so. It is now common practice to extend a chimney liner out of the top of the chimney terminal by 300mm (11¾in) and use this as the chimney pot. One manufacturer even makes a special liner for this reason without a rebate on the top but with a spigot on the bottom. Another method is to

use a tapered pot that is supported on the brickwork of the chimney. It is embedded in the chimney and tapered so that it can be sealed around the flue liner and still give the same area at the top as the flue liner.

When a chimney pot is fitted, it is often still done in the same way as it was before flue liners were compulsory. This method involves the use of four pieces of slate, one on each corner of the chimney, on which to rest the pot and stop the flaunching (mortar used to hold the pot in place) falling down the flue. **This method is not correct and does not conform to either the British Standards or the recommendations of the chimney pot manufacturers.**

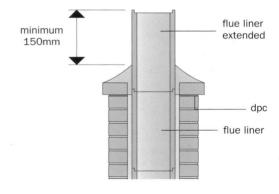

Fig 70   The final chimney liner used as a chimney pot (flue terminal).

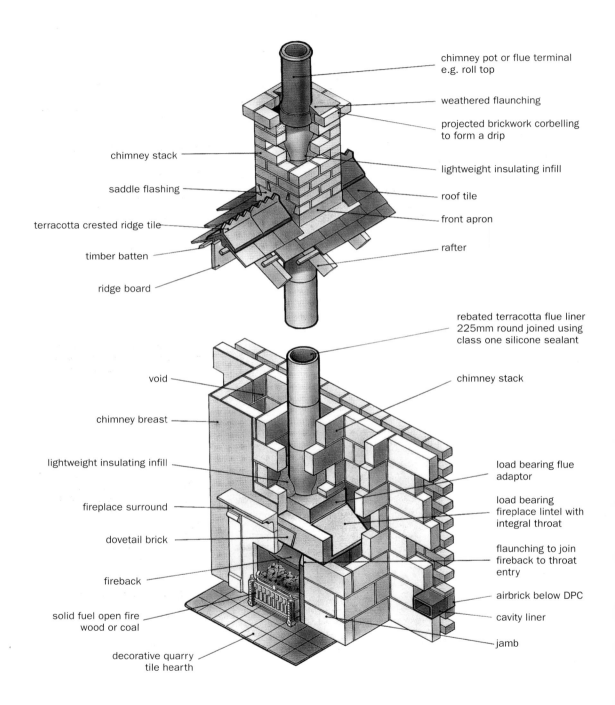

chimney pot or flue terminal e.g. roll top

weathered flaunching

projected brickwork corbelling to form a drip

chimney stack

lightweight insulating infill

saddle flashing

roof tile

terracotta crested ridge tile

front apron

timber batten

rafter

ridge board

rebated terracotta flue liner 225mm round joined using class one silicone sealant

void

chimney stack

chimney breast

lightweight insulating infill

load bearing flue adaptor

fireplace surround

load bearing fireplace lintel with integral throat

dovetail brick

flaunching to join fireback to throat entry

fireback

airbrick below DPC

solid fuel open fire wood or coal

cavity liner

jamb

decorative quarry tile hearth

Fig 71   A masonry chimney constructed to suit an open fire fuelled by either gas or solid fuel.

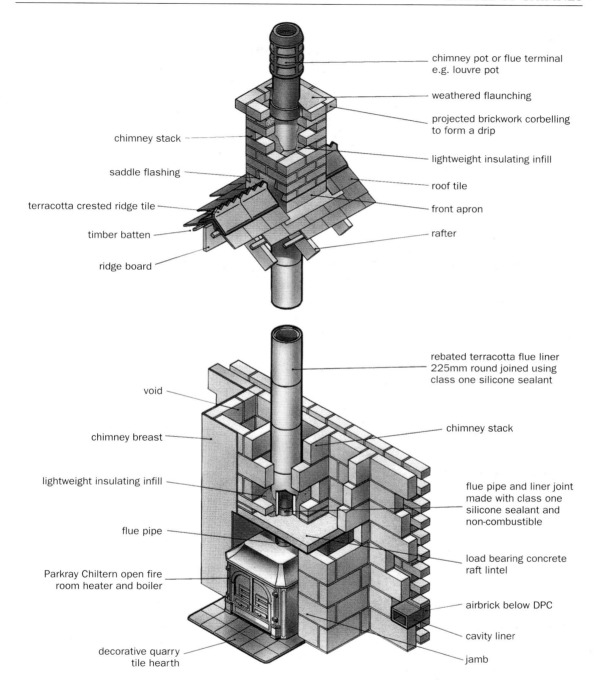

Fig 72   A masonry chimney system that has been designed and constructed with a recess and flue size that is suitable for a closed appliance with a connecting flue pipe, such as a stove.

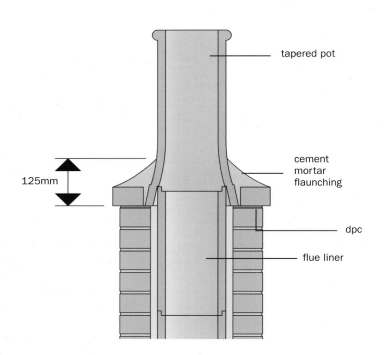

125mm

tapered pot

cement
mortar
flaunching

dpc

flue liner

*Fig 73   The use of a
tapered pot supported on
a chimney stack but
sealed onto the chimney
liner.*

The British Standards state that a chimney pot should be built into the stack to give an embedment of at least 125mm (5in), excluding any flaunching, or a quarter of the length of the chimney pot height, whichever is the greater. The British Standards go on to say that the chimney pot should be sealed to the top of the flue liner using a cement mortar. This method provides adequate stability for the chimney pot and ensures that there are no leaks into the flue right up to the point where the gases leave it.

When a masonry chimney is designed to be used with a stove, the main change is in the use of a flat reinforced lintel at the top of the fireplace recess similar to the one shown for the universal fireplace opening. This forms the base for the flue liners, and as the appliance does not require a gather none is provided. This simplifies the construction slightly, with the result that the hole in the lintel can be sized to suit the liner and the stove to be fitted.

FACTORY-MADE BLOCK SYSTEMS

The Building Regulations also include information on factory-made block chimneys and flues, and again in this section we will deal with chimney systems suitable for all fuels, as opposed to the small flue blocks that are only suitable for radiant and some live fuel effect (LFE) gas fires.

Flueblock chimney systems have to meet the same basic criteria as those for masonry chimneys, but in this case the overriding criterion for their installation is the manufacturer's installation instructions. Unlike the masonry chimney system, the acceptance criterion is the testing of the system and once this has been completed and achieved it is the system manufacturer's requirements that take precedence.

All of the systems will have differing requirements and so it is not possible to state definitively the correct way for them to be installed. However, to give

The Building Regulations *ADJ* with Reference to Factory-Made Block Systems

1.29 Flueblock chimneys should be constructed of factory-made components suitable for the intended application installed in accordance with manufacturer's instructions. Ways of meeting the requirement for solid fuel appliances (and generally suitable for other fuels) include using:

a) flueblocks whose performance is at least equal to that corresponding to the designation T450 N2 S D 3, as described in BS EN 1443: 1999, such as:

  i) clay flue blocks at least meeting the requirements for Class FB1 N2 as described in BS EN 1806:2000;
  ii) other products that are independently certified as meeting the criteria in a).

b) blocks lined in accordance with Paragraph 1.27 and independently certified as suitable for the purpose.

1.30 Joints should be sealed in accordance with the flueblock manufacturer's instructions. Bends and offsets should only be formed with matching factory-made components.

an indication of what is involved I have been given permission from one of the major producers of this type of system to use their instructions and system as an example. The system discussed here is one called the Isokern DM block chimney system.

Despite these being described as 'lightweight' concrete blocks they can be heavy to handle and so this system comes with an outer case and separate inner block to make them easier to handle. The material used here is a pumice concrete, which is based on a volcanic rock that is extremely heat-resistant. It is claimed by the manufacturer that not only will it contain a chimney fire, but it will also be able to withstand one and still be suitable for use.

Once the foundations are in place, the firechest (prefabricated fireplace recess) can be constructed or a brick fireplace recess can be created. What material is used as the top of this chest will depend on the type of appliance being installed. If an open fire is being fitted, the gather lintels will be added, which will allow a flue of 225mm (9¾in) internal diameter to be used and can provide for a fireplace of up 700mm wide × 850mm high (27 × 33in). If a closed appliance is to be fitted, then a flat reinforced raft lintel is supplied and this has a hole in the centre for the flue pipe to be sealed into.

On the top of the gather or raft lintel the first outer block is bedded (with the protruding ribs around the outside facing up) using a one-part lime, one-part OPC and six-parts sand mortar mix or the Isokern lip glue. A starter flue block is placed inside the casing, also on the bedding mix, with the rebate (or socket) facing up. If the flue is on the top of a raft lintel for use with a closed appliance, an Isokern DM adaptor should be placed under this as it will provide a spigot to which the connecting flue pipe can be sealed.

The next flue block should be placed on the starter flue block, ensuring a good fit of the spigot into the rebate. The joint should be made by filling the rebate in the lower block with Isokern lip glue and bedding the spigot on the upper liner into the rebate, pushing down to squeeze out any excess lip glue and ensuring a secure fit. The excess lip glue should be removed and the inside of the joint made as smooth as possible.

The next outer block can be added and bedded on either Isokern lip glue or mortar. It is important to ensure that the air gap between the flue blocks and the casings is not filled up or bridged while doing this. The flue blocks and outer casings are added as above in alternate steps and continued up through the room.

If an offset is required in the flue this can be achieved using the purpose-made offset blocks. These do not have an inner and outer, but are a single unit with a 30-degree angle from the vertical offset built in. They can be used in any number until the required offset is achieved, but they must be fully supported down to foundation level as they will be forced apart by the weight of the flue above them if they are not.

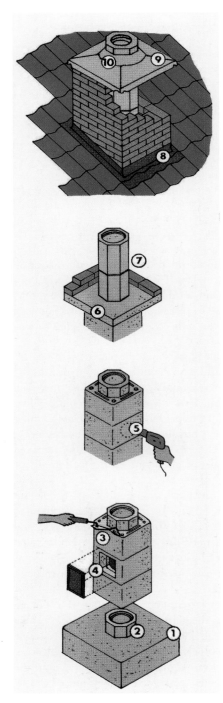

1 Construction begins by placing the first outer casing, spigot uppermost, on the foundation, lintel or gather, on a 1:1:6 cement:lime:sand bedding, or ISOKERN lip glue.

2 A starter flue block is placed inside the casing, also on the bedding, with rebate uppermost.

3 Subsequent casings and flue blocks are added, using ISOKERN lip glue to join the flue blocks, and either lip glue or mortar to join the casings. Ensure the air gap between flue blocks and casings is not filled up.

4 Soot door should preferably be situated below appliance flue entry if possible.

5 Clean-out and flue entry is achieved by either using purpose-made access flue blocks and casings, or by drilling and chiselling out access at desired point.

6 If the chimney is to be clad with bricks or stone externally, a corbel is fitted above the last casings just below roof level.

A damp proof course must be inserted in the brickwork cladding to avoid water penetration around the flue blocks.

7 If no brick or stone facing is required the externally sited DM outer casings must be weatherproofed using either waterproof render or suitable cladding.

8 Flashing for brick clad top to normal practice and standard, depending on type of cladding.

9 Purpose-made ISOKERN capping with drip rail. Alternatively, brick/concrete capping with overhang to be built 'in situ'.

10 Top of flue block (or liner) and capping flaunched with mortar. Fit pot if required for aesthetic reasons.

The last flue block below the offset will have to be cut in length so that it is level with the top of the last outer casing below the offset. When the offset is complete, the first flue block above the offset should be another half-height starter flue block and then the staggered joint system is started again and the flue blocks and outer casings are added as before.

Where the chimney system reaches a structural floor or ceiling, the structural timbers must be trimmed back to achieve an air space of at least 30mm (1¼in), but non-structural woodwork such as skirting and floorboards can be taken directly up to the outer casing wall. To maintain the 30mm space, and to provide some lateral location for the chimney system, a non-combustible spacer such as mineral board should be secured in place between the trimmed joists and the outer casing wall. This only needs to be a spacer and does not need to fill the space completely.

Fig 74 A series of illustrations from the Isokern DM system installation instructions.

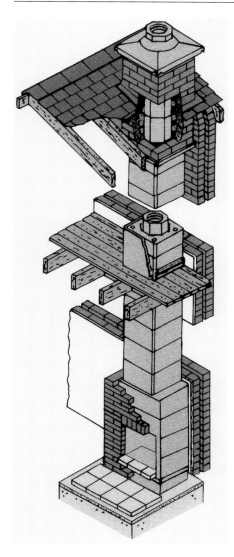

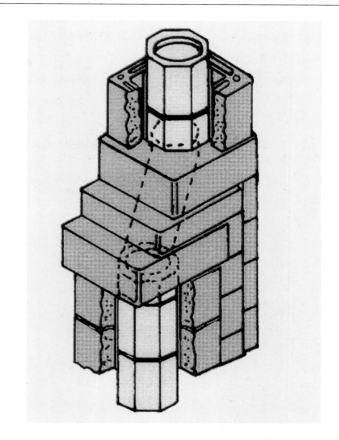

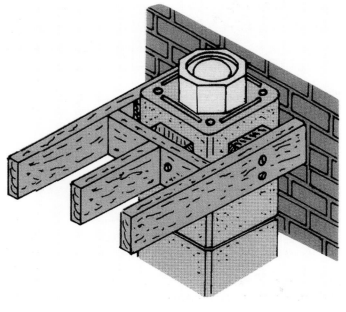

ABOVE: *Fig 75   Isokern DM chimney systems forming an internal chimney for an open fire.*

TOP RIGHT: *Fig 76   Forming an offset using specialist blocks from Isokern.*

RIGHT: *Fig 77   Penetrating a floor with the DM system requires the trimming of floor joists and the fitting of non-combustible spacers to provide lateral support.*

Fig 78   The manor firechest by Isokern for use with the DM chimney system providing a fireplace opening of up to 1,150mm wide × 810mm high.

Again, the flue continues upwards until it is just below the roof surface and at this point there are two options:

- The flue can continue up out of the roof and have flashing and damp-proofing placed in as the construction continues upwards. If this option is chosen, the outer casing has to be coated with a waterproof render or similar weatherproof cladding.

- A corbel block can be added and the flue block continued upwards. The outer casing is replaced at this point with either a brick or stone cladding, which is flashed into the roof and damp-proofed as normal. If this is done, the space between the flue block and the brick or stone cladding should be filled with an insulating mix of twenty parts leca and one part OPC with a small amount of water to just wet the mix.

British Standards Referring to Prefabricated Stainless Steel Systems		
Fuel Type	**Product Standards****	**Installation Standards****
Gas*	BS 715	BS 5440: 2000
Oil*	BS 4543 Part 3	BS 7566 Parts 1–4***
Solid fuel (including wood)	BS 4543 Part 2	BS 7566 Parts 1–4***

Notes
* Flues to BS 4543 Part 2, installed in accordance with BS 7566 Parts 1–4, can be used with all fuels for non-condensing appliances.
** The most recent issue of any Standards must be used.
*** To be replaced by BS EN 12391.

Isokern makes a capping block for the chimney for both the outer block and brick-clad chimney stacks, that includes the hole for the last liner to protrude through and act like a chimney pot. The advantage of using the flue block as a chimney pot in this case is that it is warm and insulating in its own right, and so will not have the same cooling effect as a tall clay pot.

There are many other extras to the system, such as purpose-made soot door casings and flue blocks, as well as breach blocks to allow the flue pipe from a free-standing or rear-exit stove to enter the flue. One important factor about this system is that it can be specified in larger sizes, up to 300mm (11¾in) internal diameter with a specialist firechest that can allow for an opening size of up to 1,150mm wide × 810mm high (45 × 31½in). Since all the components are designed to work together this greatly reduces the risk of failure when dealing with such a large fireplace.

PREFABRICATED STAINLESS STEEL SYSTEMS

This type of chimney comes in component parts and can be put together without the need for specialist skills such as bricklaying, welding or braising, but the finished product can be equally as dangerous as other systems if the manufacturer's installation instructions have not been carefully followed. Again, as there are a lot of manufacturers of these systems the information given in the Building Regulations is of limited use.

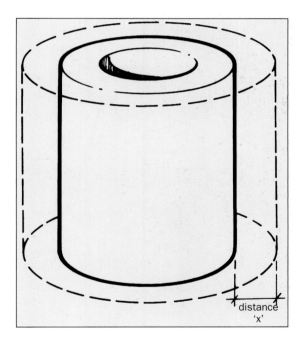

Fig 79   The distance 'x' for clearance from combustible material. The distance represented by 'x' is given in the manufacturer's installation instructions and the relevant British Standards.

The *ADJ* gives information on the British Standards with which the flues should comply for the various fuels, as well as the British Standards that cover their installation.

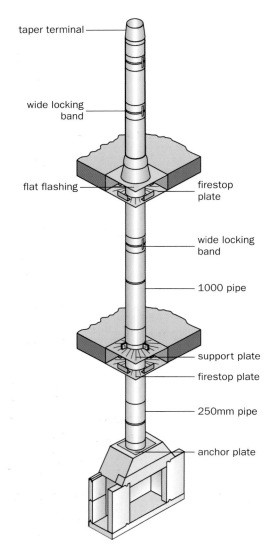

taper terminal

wide locking band

flat flashing

firestop plate

wide locking band

1000 pipe

support plate

firestop plate

250mm pipe

anchor plate

Fig 80   A prefabricated chimney system incorporating a gather for an open fire which provides support.

In addition to this information, some specific requirements are made in the Regulations to supplement the Standards given above and these include the following:

- Where a factory-made metal chimney passes through a wall, sleeves should be provided to prevent damage to the flue or building through thermal expansion. To facilitate the checking of gas-tightness, joints between chimney sections should not be concealed within ceiling joist spaces or within the thicknesses of walls.

- When providing a factory-made metal chimney, provision should be made for it to be possible to withdraw the appliance without the need to dismantle the chimney.

- Factory-made metal chimneys should be kept a suitable distance away from combustible materials by locating the chimney not less than distance 'x' from combustible material, where 'x' is defined in BS 4543–1: 1990 (1996).

- Where a chimney passes through a cupboard, storage space or roof space, a guard should be provided that is placed no closer to the outer wall of the chimney than distance 'x' above.

- Where a factory-made metal chimney penetrates a fire compartment wall or floor, it must not breach the fire-separation requirements as given in *Approved Document B* on fire safety. These requirements may be met by using a factory-made metal chimney of the appropriate level of fire resistance, or by casing the chimney in non-combustible material, giving at least half of the fire resistance recommended for the fire compartment wall or floor.

Since a British Standard exists for the installation of these types of chimneys the methods and principles do not change very much from one make to another; just the components will vary. I have been given permission to reproduce elements of diagrams from the installation instructions for the Rite-Vent products and these form the basis of this description.

The flues that can be used for gas appliances are of a much lighter construction and of materials that have a lower level of corrosion resistance as there is no risk of a chimney fire and the condensates are less acidic. These flues are lightweight and easier to handle, but have to be treated in much the same way as a flue for oil or solid fuel, so it is these products we will consider here.

The illustrations here show three different types of installation: one for an open fire with a prefabricated chamber forming the fireplace recess and supporting

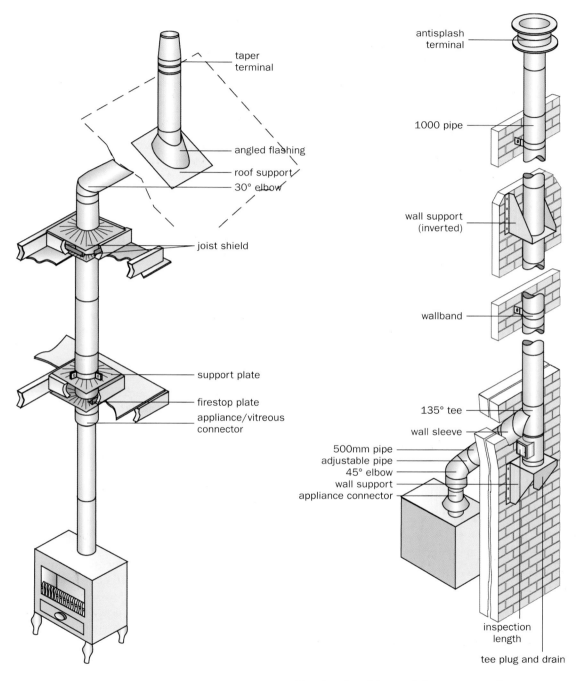

Fig 81   A prefabricated chimney system for a free-standing stove supported at ceiling level.

Fig 82   A prefabricated chimney system for a free-standing appliance, passing through an outside wall and supported externally.

89

the flue; another where the appliance is not in a recess and the chimney system is supported at the ceiling above; and a third showing the chimney system passing through a wall, being supported by brackets and so on.

In the first instance, some consideration has to be given to the floor that the flue will stand on and whether a change to the foundations will be required. If it is a solid floor then this is not normally necessary, but it is best to get the opinion of a surveyor.

It is important to conform to *Approved Document J* when installing this type of chimney, as it is with all the types of chimney previously discussed. When designing the flue system there are a number of things to consider in addition to those already discussed. Firstly, the clearances from combustible material need to be taken into account and the effect this may have on parts of the construction such as floor joists, Gang-Nail trusses and other roof structures, support required for exposed lengths, locations of brackets, internal supports and fire stops, the angle and lengths of offsets and so on. Any of the above may dictate the final location of the fireplace or appliance.

The first step is, as before, to decide on the diameter of the flue to be used. These systems are available from 125mm (5in) up to 600mm (23in), but the most common sizes are 150mm (6in) to 250mm (9¾in) and these sizes are the most economical to use. There is a large range of components in each size and these must be chosen carefully. They must always be used as described by the manufacturer and must never be modified, as this could make the component or the whole chimney dangerous; nor would the chimney conform to the required Standards and Regulations if components have been modified.

The selection of section lengths and so on must be made to ensure that there are no joints in the system between ceilings and floors. Protection such as boxing-in must be considered, as this is required whenever the flue passes through any room, cupboard or space other than the one in which the appliance is situated. Any such enclosure must be constructed in such a way as to provide the required fire protection and still allow access to the chimney system for inspection and maintenance.

In some cases, the manufacturers restrict the angle of bends and offsets in the smaller sizes to only 30 degrees from the vertical – this should be checked at the design stage. Where an offset is required, other than where the flue pipe joins the chimney system, this should not exceed 20 per cent of the overall effective height of the flue and only one such offset should be used.

Consider the load-bearing requirements of the chimney system and which component will be used where. Most manufacturers give a table of load-bearing data for the relevant components, as well as information on whether they should be used internally or externally. As a rule of thumb, the following information could be considered but must be checked with the manufacturer's data before installation commences.

The connection between the appliance and chimney system can either be direct, using the required adaptor, or indirect, using a flue pipe connection and the correct adaptor for the type of flue pipe in use. Any flue pipe connection to the chimney must be sealed and made in the same room as the appliance.

Where a chimney is required to extend more than 1.5m (5ft) above the surface of the roof it will normally require additional support such as guy wires, a ladder or mast supports. The type of support required and the exact components required will vary according to the manufacturer, but some types are discussed below.

Chimney Supports

Where a prefabricated chimney is to be fitted on top of a recess or gather unit the base of the flue is normally supported by this, and located and secured using an anchor plate. However, where the appliance using the flue is not in a recess but is free-standing then another type of support is required.

This support will vary but under no circumstances should the appliance or flue pipe take any of the weight of the chimney system as it should be possible to remove the appliance and any connecting flue pipe without disturbing the chimney system.

Telescopic Floor Support
This component is designed to provide support for

the flue directly from the floor. To do this it will have to be used with either a 90-degree tee (only when connected directly to the rear outlet of the appliance), or a 45-degree tee. This is the strongest of the support components readily available and in most cases can support up to 16m (53ft) of chimney system.

Wall Support

These can usually be used internally or externally and can be used at the base of a chimney, supporting a tee, or intermediately within the chimney system. These units can be fitted with the side plates either underneath the base plate or with them above it. If the side plates are under the base plate the wall support can support up to 9m (30ft), while it can support up to 15m (49ft) with them above.

Telescopic Ceiling Support

This is a support system used where the chimney system passes through the ceiling of a room. In addition to the support role this component has, it also spaces the chimney from combustible material, provides an additional heat shield and provides the fire-stop rating of the floor it passes through (this avoids there being a weak point in the floor above for fire protection).

This component will normally be capable of supporting at least 6m (19ft 9in) of flue, of which some can be suspended below the support and the remainder above it.

Adjustable Roof Support

This component should be used to provide lateral stability as well as load-bearing support at the point where the chimney system passes through the roof surface. The unit can provide lateral support for up to 1.5m (5ft) of flue above the roof surface and up to 9m (29½ft) of flue, of which up to 6m (19¾ft) may be suspended below it. The brackets are normally adjustable to suit all angles of roof pitch, including flat roofs.

Lateral Stability

As well as providing support for the weight of the chimney system, all of the above components provide varying degrees of lateral support. Despite this, most chimneys of this type will require some additional lateral support and the following components are designed to provide this. In addition to the load-bearing components, some kind of lateral support component should be provided at least every 2.5m (8ft).

Wall Bands

There are a number of different designs of wall band – some are for internal use only, while others can be used both internally and externally. A number of manufacturers make a high-stability wall band, which, although it still needs to be used every 2.5m (8ft), will provide additional stability in exposed locations where the wind may achieve higher forces than normal. At least one manufacturer makes an adjustable wall band that can stand the chimney anything between 50–300mm (2–11¾in) away from the wall supporting it.

CHAPTER 7

Using an Existing Chimney

SURVEYING AND TESTING

Many people who consider the installation of a fireplace, fire or stove are doing so because they already have a chimney in the property and wish to change its use, or, if it has been closed off, to bring it back into use.

This makes sense in these days of the sterile central heating radiator, as a way of returning the soul to a room or the property as a whole. However, one thing we must remember is that the chimney we are about to use will have been constructed to meet the needs of the time, and it may well have been used for years, in some cases hundreds of years, since it was built. Over these years, the types of appliances have changed and also use of the chimney may have resulted in damage or a reduced life expectancy. For these reasons, reopening or changing the use of a chimney is a more risky business than constructing a new one. Despite this, if the chimney is adequately surveyed, tested and, if required, adapted to suit its new use, then the additional risks can be eliminated. *Approved Document J* now states that bringing a flue back into use or changing the use of a flue requires the person doing the work to carry out certain tests to ensure the suitability of the chimney for the purpose proposed.

The survey and tests required are: (a) visual inspection; (b) chimney-sweeping check; (c) core ball test; (d) chimney air-tightness test; and (e) smoke evacuation or spillage test. The visual inspection is run through in this chapter, while the other tests are detailed in Chapter 10. It is important to realize that skill in interpreting what is seen and the results of the tests is important and can only be gained by experi-

ence, so the certification of a chimney should be carried out by a chimney professional.

VISUAL INSPECTION

- What size is the flue in the chimney and is it suitable for the application intended (not too small or too large)?
- Does the external appearance of the chimney seem to be in good condition? For example, are there any cracks in the brickwork or rendering, are the mortar joints disintegrating or receding and is there any external staining? This check should include all visible parts of the flue such as on a gable end, on the chimney stack as well as in the roof space.
- Does the chimney stack appear to be vertical and square, or is it leaning and bowed in places?
- Do the brickwork joints need pointing, particularly on the chimney stack, and is the flaunching at the base of the pot in good condition, or is it cracked, weak and soft?
- Is the chimney pot in good condition, clear of deposits and of the correct size and type for the appliance to be installed?
- Are the weather flashings between the roof and chimney stack in good condition or are they torn and loose? Is there a damp-proof course or lead tray in the chimney stack?
- Does the chimney terminate in a location that complies with the Building Regulations (paying attention as to whether the roof surface could be considered easily ignitable)? Is there any issue that may impair the draught in some weather

conditions, such as tall buildings or trees close by, or is the property in a deep, steep-sided valley or on the side of a hill?

- Is the chimney of sufficient height to ensure a reasonable updraught? Does it meet the 4.5m (15ft) minimum suggested in the Building Regulations?
- Are there any openings, vents or grills (inside or outside the property) in the chimney that might spoil the updraught and is there only one fireplace served by the flue?
- Are there acute changes in direction in the flue where deposits might settle and not be effectively cleared during sweeping? Are there any long offsets that will add considerably to the resistance in the flue?
- Is there a large fireplace recess that will form a cavity at the base of the flue when the new appliance is installed, allowing a build-up of soot that will add to the risk of a chimney fire?
- Is there a gather or a flat lintel at the top of the fireplace recess? If there is a gather is it at least 45 degrees from the horizontal and smooth, or is it steep and rough corbelled?
- Has the chimney been swept recently?
- Is there adequate, permanently open, purpose-provided ventilation to the room for the new appliance?

In addition to these observations, if it is possible to discuss the situation with someone who has seen the chimney in use then there are some useful questions you can ask:

- Has there ever been smoke or fumes coming into the room at any time? If so, has this been continuous or intermittent and only in certain weather conditions?
- Has there ever been the smell of smoke or fumes in any other part of the house?
- Can it be difficult to light the appliance or get the fuel to burn up quickly?
- When was the chimney last swept?
- What type of fuel was used?
- Has there ever been a chimney fire?
- Are there any signs of dampness or staining on the chimney wall either inside or out?
- When was the last time the chimney was in use?

All of the above will provide useful information to help to establish if the chimney was stopped from being used because of a fault or just because the use of a fire fell out of fashion.

The tests described in Chapter 10 should be carried out at this stage if an existing chimney is to be used, as this will reduce the testing needed at a later stage, as well as ensure that any faults and remedial works are identified at this early stage. It can be very frustrating to go through the exercise of fitting a new fireplace or appliance only to find that the chimney is damaged and needs a lot of money spent on it to make the new fire work safely and correctly.

The tests that should be carried out are: sweeping check; core ball test; and air-tightness test. If there is already an appliance in place and it is possible to carry out a smoke evacuation test, then this would also be advisable considering the small amount of extra cost involved.

CHAPTER 8

Refurbishment and Repair of Flues

The refurbishment, or repair, of a chimney can entail anything from the rebedding of the chimney pot to the full relining of the flue. It is important to identify the required repair in order to carry out only the required work, but ensure all the required work is carried out.

Even simple tasks such as bedding a chimney pot or pointing the chimney stack have a right and a wrong way to be done, and many of the tasks that may be required are explained here. However, remember that these are only explanations to ensure that you are aware of the work involved and they are not intended as a full step-by-step guide. As with the construction of flues, the work is specialist in nature and so should be left to the professional.

It is important that any work at chimney terminal level is carried out using safe roof-access methods and working platforms after a full risk assessment of the work has been carried out by a trained and competent person who is familiar with the risks and regulations involved when working at heights.

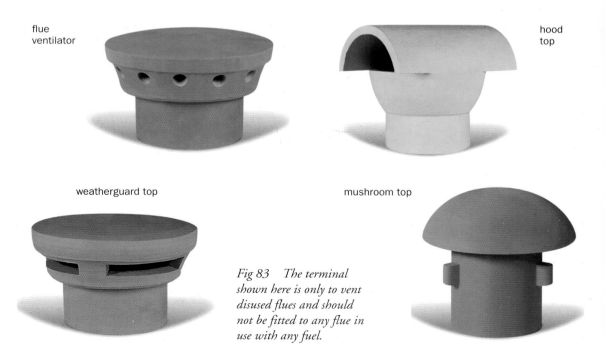

flue ventilator

hood top

weatherguard top

mushroom top

Fig 83   The terminal shown here is only to vent disused flues and should not be fitted to any flue in use with any fuel.

94

Fig 84   A flue terminal guide for quick reference to fuel type and common terminals.

	Pot title	Suitable for	Not suitable for
	Rectangular based octagonal pot	May be used for solid fuel closed appliances, gas and oil fired appliances dependent on outlet size	Not suitable for solid fuel open fires unless outlet of pot is 200mm internal diameter minimum
	Beaded rebated square pot	Solid fuel open fires and closed appliances depending on outlet size, gas and oil appliances	
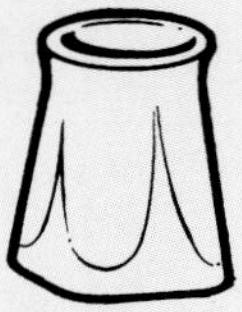	BSS type 4D – square to round pot	Suitable for all types of appliances subject to correct outlet diameter	If less than 200mm outlet diameter, not suitable for solid fuel open fires
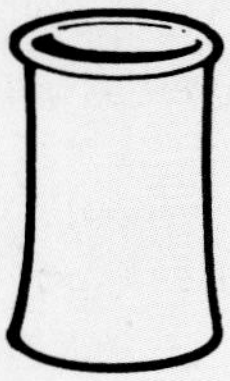	BSS type 5E – tapered round pot	Suitable for all types of appliances subject to correct outlet diameter	If less than 200mm outlet diameter, not suitable for solid fuel open fires
	Square plain pot	Suitable for all types of appliances subject to correct outlet size	If less than 185mm square outlet, not suitable for solid fuel open fires
	Square spiked pot	Suitable for all types of appliances subject to correct outlet size	If less than 185mm square outlet, not suitable for solid fuel open fires
	Louvre pot	Suitable for all types of appliances subject to correct outlet diameter	If less than 200mm outlet diameter, not suitable for solid fuel open fires

95

Fig 84   (continued)

	Pot title	Suitable for	Not suitable for
	Marchone flanged base	This is a traditional pot to help prevent turbulent downdraught but will also reduce rain ingress. The top outlet is less than the 200mm internal diameter but the total outlet area is still sufficient for a solid fuel open fire	Not used with gas appliances other than DFE fires
	Round-based bishop	For use with solid fuel open and closed appliances and oil appliances	
	Weatherguard top	Venting a flue only	Not suitable for any appliance or fuel
	Flue ventilator	Venting a flue only	Not suitable for any appliance or fuel
	Hood top	Venting a flue only	Not suitable for any appliance or fuel
	Bishop GC2	May be used with all gas appliances with flue outlets up to 125mm internal diameter and some LFE gas fires which are rated as suitable for use with class two flue blocks	Not suitable for solid fuel, oil or gas DFE appliances
	GC5 insert	May be used with all gas appliances with flue outlets up to 125mm internal diameter and some LFE gas fires which are rated as suitable for use with class two flue blocks	Not suitable for solid fuel, oil or gas DFE appliances
	DFE pots	Suitable for solid fuel open and closed appliances, gas DFE and LFE fires requiring a minimum 175mm internal diameter flue	Not suitable for use with class two flue blocks or where a gas flueway has been reduced to less than 175mm in diameter

BEDDING A CHIMNEY POT

A chimney pot is an integral part of any masonry chimney system, as is the terminal of a prefabricated system. It is important to recognize that each of the chimney terminals illustrated here is suitable for specific flues and appliances only.

The type of appliance will dictate the type of terminal that can be used. Provided here is a quick reference guide to the type of pot that can or must be used with certain types of appliances. When fitting a chimney pot, it must be secure, sealed to the flue and waterproof to prevent excess rain entering the chimney.

It is often not recognized that a chimney pot should not be set directly on top of the brickwork of a chimney stack, but should be bedded into the top of the stack to ensure it has adequate stability in high winds. Some chimney pots are well over a metre tall and can weigh up to 100kg (220lb), and so will cause considerable damage if they fall from the top of the chimney stack.

All chimney pots must be bedded at least on a course of bricks, or 25 per cent of their total length must be located down in the chimney stack. If the chimney is lined, the chimney pot must be sealed directly onto the top of the chimney lining, other than in the case of a relining with stainless steel, when the lining must pass through the chimney pot.

It is common practice to place pieces of slate across the corners of square flues and rest a round chimney pot on top of these. The slate is intended to support the pot and prevent the mortar flaunching from falling down the gaps that would be formed in the corners. There are several problems associated with this, including the sudden change in area between the square flue and the round chimney pot, the fact that the chimney pot is not set into the chimney stack, and that the slate will often protrude into the base of the pot, creating an even greater restriction. A chimney pot on a square flue, therefore, should have a square base. It should be at least the same internal dimensions as the flue so that it can be seated directly onto the brickwork. At least one course of bricks must be placed around the chimney pot so as to provide protection and stability.

The flaunching around the pot should consist of one part sulphate-resisting cement and half a part

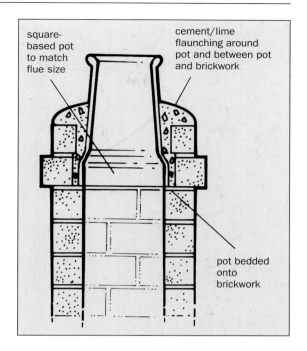

Fig 85   The correct fitting of a chimney pot onto an unlined square flue.

hydrated lime to four and a half parts of sand. It is advisable to include an additive to provide permanent frost resistance. If the existing mortar is weak then additional additives are available, such as styrene-acrylic-based compounds, to improve the adhesion to existing bricks and mortar. The flaunching should form a concave shape to avoid any thin edges that will be vulnerable to frost and damage.

When a round chimney pot is fitted on top of a round flue, the inside diameter of the pot should be at least equal to that of the flue and it should be parallel-sided to ensure that it does not cause a restriction at the top. If a tapered pot is to be used, the base of this should be large enough internally to pass over the outside of the liner and the terminal diameter should be at least equal to that of the flue, again so as not to cause a restriction. It is common practice these days to pass the last liner up out of the stack and to use this as a chimney pot; some liner manufacturers provide a special 'last liner' that does

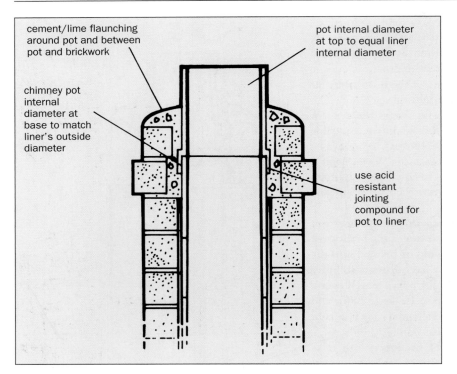

cement/lime flaunching around pot and between pot and brickwork

pot internal diameter at top to equal liner internal diameter

chimney pot internal diameter at base to match liner's outside diameter

use acid resistant jointing compound for pot to liner

Fig 86 A straight-sided round pot on a round flue liner.

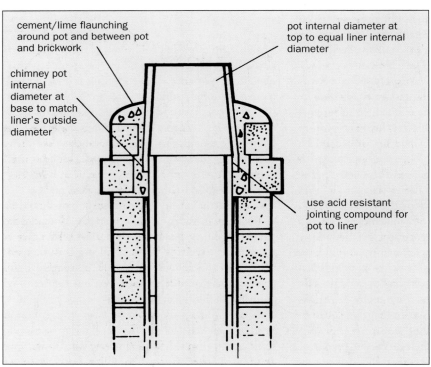

cement/lime flaunching around pot and between pot and brickwork

pot internal diameter at top to equal liner internal diameter

chimney pot internal diameter at base to match liner's outside diameter

use acid resistant jointing compound for pot to liner

Fig 87 A tapered pot fitted over a round flue liner.

not have a rebate in the top so as to provide a neater finish.

Note that all of these types of installation result in what is known as an oversail or drip course. This partly corbelled out-course of bricks creates an undercut that the rain and water will not follow. This means that the dirty and acid water that will run down the outside of the chimney pot will run over this course and then drip off onto the tiled roof, where it will be diluted by the rest of the rain on the roof and wash away down the gutters without causing damage. If this course is not present, the acid water will run down the surface courses of the chimney stack, allowing the acid to damage the mortar in the joints.

REPOINTING A CHIMNEY STACK

Repointing is required when the mortar in the stack has receded 10mm (⅓in) or more into the joint or has become cracked, crumbly or loose.

The repointing of a chimney stack should be carried out in a strict order, because otherwise it can become destabilized while the work is being carried out. The sequence shown here should be followed. The actual pointing should be of the 'bucket-handle profile' type, as this will help to prevent acid attack from the rain that will have absorbed soots and condensates and then run down the outside of the stack.

The first part of repointing is to rake out the damaged mortar until sound material is reached, creating a square profile. If the existing mortar is sound below the surface a depth of at least 25mm (1in) should be adequate, but if it is crumbly or cracked a depth of 35mm (1⅓in) will be required.

This raking out and then repointing should be done one face at a time, with at least two days being left before work on the next face in the sequence is commenced.

The recommended mortar mix is one part sulphate-resisting cement, half a part hydrated lime to four and a half parts of sand. It is advisable to include an air-entraining additive to provide permanent frost resistance. If the existing mortar is weak additional additives are available, such as

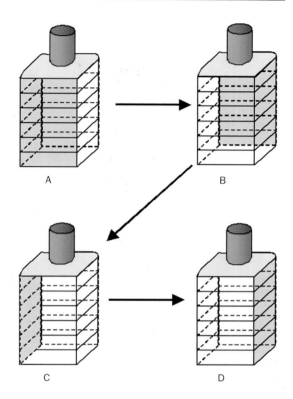

Fig 88   *The correct sequence for repointing a damaged chimney stack.*

styrene-acrylic-based compounds, to improve the adhesion to existing bricks and mortar. If a repointing gun containing premixed mortar is used, the mix must be suitable for use on chimney stacks (it must have the required acid resistance).

Once the raking out of the face to be worked on has been completed, the masonry should be cleaned off and dampened. The new mortar must be pressed firmly into the joint to avoid trapping any air and then should be finished with a 'bucket-handle profile' to minimize rain penetration.

REPAIRS TO BRICKWORK

With care, localized brick deterioration can be repaired. If the amount of spalling (the loss of the

REBUILDING A CHIMNEY STACK

Visibly leaning stacks and those with cracked or seriously damaged masonry will need to be rebuilt. It is very important that if any part of the chimney stack has a wall that forms part of any adjoining property that a party wall agreement is put in place prior to work commencing. For more details on this, contact a legal advisor, surveyor or the local building control office.

The materials used in the rebuilding of a chimney stack should be as follows:

- Bricks – Type F, BS3921, for clay, or Types 3 to 7, BS 187, for calcium silicate. Mortar – one part sulphate-resisting cement, half a part hydrated lime to four and a half parts of sand (it is advisable to include an air-entraining additive to provide permanent frost resistance).
- Flashing and damp-proof courses – code four lead (code five may be specified in very exposed positions and for some details.
- Liners must be of the non-metallic types and match the existing ones if present (if no chimney lining is present in the original chimney and it is not anticipated that the whole flue will be lined, it is best not to line the new stack as this will cause a change in shape and resistance part way up the flue).

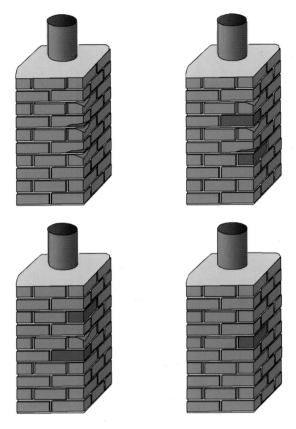

Fig 89   Replacing damaged bricks in a chimney stack without the need for a rebuild.

It is important that a decision is made as to whether the property is considered to be sheltered, exposed or severely exposed, as this will make a difference to both the material and methods used for damp-proofing and flashing. If you are unable to identify which of the above categories the property falls into, consult the local planning department or obtain advice from the Chartered Institution of Building Service Engineers Guides.

The chimney stack should be dismantled until sound brickwork is reached; this is likely to be two to three courses below the roof level at the lowest point. The rebuilding work should be carried out using new bricks as specified above – the reuse of bricks from deteriorated masonry is not recommended. The chimney stack is then built up to the level of the first damp-proof course (dpc). For shallow pitched roofs

brick face or corners) or deterioration is not too widespread, bricks can be stitched in to restore strength and stability. The existing damaged bricks can be cut out with a mechanical cutter, but this must be done in stages over time so as to retain the stability of the chimney stack. No more than one brick should be removed from a course at a time; single bricks in different courses can be removed and replaced at the same time provided there are two clear courses between the bricks being removed.

The replacement bricks should match the existing ones as much as possible and be frost-resistant (Type F, BS 3921, for clay or Types 3 to 7, BS 187, for calcium silicate. Any clay bricks should have a low sulphate content and the mortar mix used should be the same as for repointing).

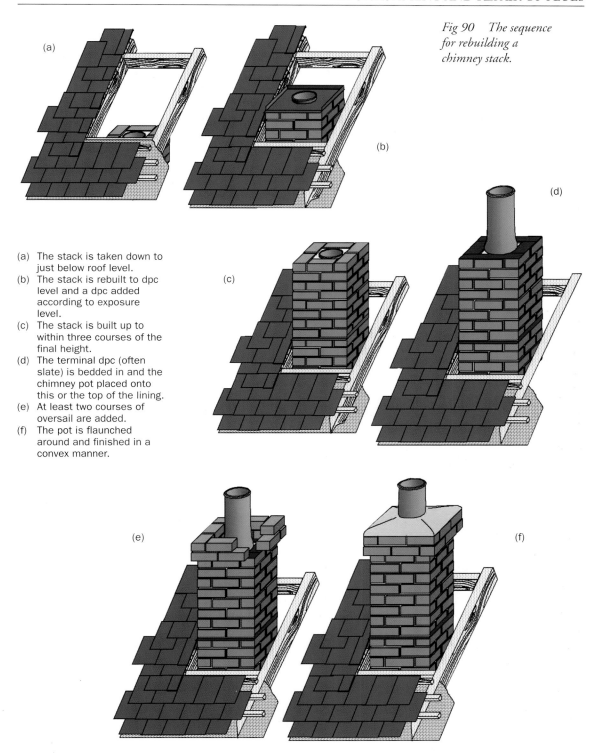

Fig 90   The sequence for rebuilding a chimney stack.

(a) The stack is taken down to just below roof level.
(b) The stack is rebuilt to dpc level and a dpc added according to exposure level.
(c) The stack is built up to within three courses of the final height.
(d) The terminal dpc (often slate) is bedded in and the chimney pot placed onto this or the top of the lining.
(e) At least two courses of oversail are added.
(f) The pot is flaunched around and finished in a convex manner.

the dpc is normally positioned just above the front apron (this is the most forward part of the flashings that will be added later); for steeper pitched roofs the dpc should be level with the back gutter. Two dpcs may be needed in very exposed positions.

The dpc is best built as a tray with at least 25mm (1in) upstands both inside and outside the chimney and a slightly larger downfold at the front to dress over the apron to be fitted later. The tray must be coated on both sides with solvent-based bituminous paint before installation, the underside of the dpc downfold should not be painted at this stage.

The way in which the upstands inside the chimney stack are dressed will depend on the flue construction and the level of exposure that the chimney is subject to, but the following are normal rules of thumb:

- If there is no liner in the flue then whatever the exposure level, the upstands must be dressed directly up against the inside of the brickwork forming the chimney stack. This dressing should be as neat as possible and tight up against the brickwork to avoid causing a restriction in the flue and to prevent it being damaged by the sweep's brush and rods over the coming years.
- If a sectional lining, such as terracotta, concrete or pumice, is present and the chimney is rated as sheltered or exposed, the dpc should not penetrate the flue lining and the upstands should be around the outside of the liner.
- If a sectional lining, such as terracotta, concrete or pumice, is present and the chimney is rated as very exposed, the dpc should be lined up with a joint in the lining and should pass through the joint and the upstands should be dressed directly up against the inside of the lining. Again, this dressing should be as neat as possible and tight up against the inside of the lining to avoid causing a restriction in the flue and to prevent it being damaged by the sweep's brush and rods over the coming years. If the pitch of the roof is steep, it is often better to place a second damp-proof tray in the stack. The lower tray should be at the point where the front apron will be and the second tray just above the back gutter level. If necessary, the outside faces of the stack can be dressed with lead between the two levels.

The brickwork should then be rebuilt to the point where the oversail or drip course is to be constructed, omitting pointing where flashings will be required. It is important that care is taken to ensure that when the chimney stack is completely rebuilt it will terminate in the correct position as dictated by diagram 2.1 in *Approved Document J* of the Building Regulations.

At this point, the terminal dpc should be bedded in; often slate is used. The pot should be sat either on top of this if no lining exists, or bedded and sealed directly on top of the lining. The base of the pot should be the same size, and importantly shape, as the top of the flue or lining. If an unlined square flue is present, the base of the pot should be square, and if the lining is round then the base of the pot should be round.

The oversail course should be constructed of at least two layers of engineering brick and then the space between the pot and these courses should be filled with a mix of lime cement as specified earlier. The same mix should then be used to flaunch over the top of the stack and around the pot. The flaunching should be concave, not convex, on top of the stack to ensure that there are no very thin sections and to provide better support and protection for the chimney pot.

The flashings then need to be added and how this is done will depend on exposure levels, steepness of roof pitch and the location on the roof of the stack.

RELINING EXISTING FLUES

There are a number of ways to reline or internally repair a flue in a chimney stack, and each method is considered in some detail below. Before we consider the installation of each type it is important to look at their respective advantages and disadvantages, and to study when and where they should be used.

Some of the methods discussed are suitable for all types of appliances and fuel types, while others are restricted to certain applications, so an overview is given here of what method can be used with which fuels and appliances.

Flexible Stainless Steel Liners

This is the most popular relining system in the UK

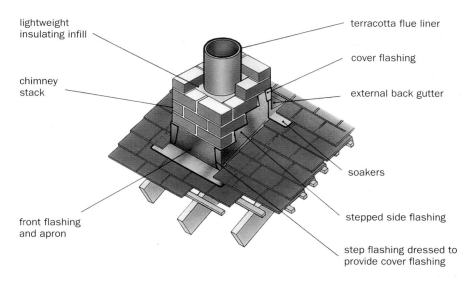

lightweight insulating infill

terracotta flue liner

chimney stack

cover flashing

external back gutter

soakers

stepped side flashing

front flashing and apron

step flashing dressed to provide cover flashing

Fig 91   Various flashing arrangements for chimney terminals.

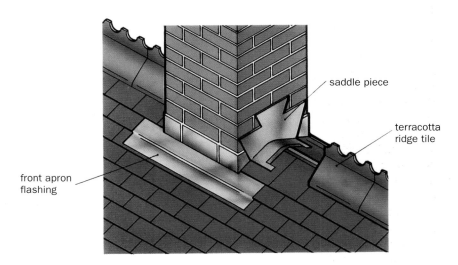

saddle piece

terracotta ridge tile

front apron flashing

and is a one-piece flexible stainless steel liner. It comes on roles of up to 30m in length and is cut to length to suit the flue to be lined.

This type of lining has become very popular as it is quick to install and often involves less disruption to the property. There are some disadvantages to the system and some limitations to its use, especially when being fitted to a solid fuel appliance where it is intended to burn smokeless fuel slowly, such as a central heating boiler.

Advantages

- Have no joints to leak other than at the top and bottom of the flue.
- Have a fixed internal diameter, reducing the risk of reductions in size throughout its length.
- Can often be fitted without disturbance to decoration.
- Quick and simple to install.
- Does not require specific skills or equipment.
- Often the least expensive alternative.

103

Flue Liner and Fuel Suitability Table					
Lining type	**GAS**	**OIL**	**WOOD**	**MULTIFLUE**	**SMOKELESS**
Pumice sectional Isokern Anki Marflex etc.	✓	✓	✓	✓	✓
Concrete Dunbrick Redbank Thurrock etc.	✓	✓	✓	✓	✓
Clay/Terracotta Hepworth Redbank etc.	✓	✓	✓	✓	✓
Cast in-situ pumped perlite NACE CICO	✓	✓	✓	✓	✓
Thin wall coatings Isokoat Eldfast	✓	✓	✓	✓	✓
Rigid Stainless Steel Ritevent Prima Plus	✓	✓	✓	✓	✓
Twin Wall Flexible Stainless Steel** Chimflex Multiflex Selflex Fireflexmaster etc.	✓	✓	✓	✓	✓*
Single wall flexible stainless steel**	✓***	✓			

*   not suitable where the appliance will be used at low output for prolonged periods
**   Mainly used to adjust a flue not repair it, as the flue must be structurally sound before insertion of the liner
***   Must be rated at T300 or greater for DFE and other gas fires

- Have a small thermal mass (they will warm up quickly and cool down quickly), so very little condensation should form when the fire is first lit.

Disadvantages
- Stainless steel liners are a temporary repair only, which means that every time the appliance in use changes, the existing lining should be removed and a new flue lining installed. According to *Approved Document J* of the Building Regulations: '1.39 If a chimney has been relined in the past using a metal lining system and the appliance is being replaced, the metal liner should also be replaced unless the metal liner can be proven to be recently installed and can be seen to be in good condition.'
- These flues have a limited life expectancy, in some cases only ten years, and this will be very dependent on the use and maintenance of the appliance and flue.

Important note: If solid smokeless fuel is to be used and there is a risk of long periods of slow burning, then the life expectancy of these liners

will be significantly reduced, in some cases to less than five years. If there is a risk that these conditions may apply, a non-metallic liner system should be used.

- If these liners become coated in tar or blocked, they are likely to be damaged during the cleaning or clearing operation as they are relatively delicate.
- Nearly all manufacturers state that the liner should only be used in a sound flue and chimney stack; for this reason if the flue is found to leak badly or have structural problems these liners should not be used until major repairs have been carried out.
- These liners are not intended to serve as a cheap form of repair to a damaged flue or to be used to change its use.
- These liners must be replaced after a chimney fire.

Flexible stainless steel liners come in two basic types and the fuel to be used will dictate the grade of liner that is needed.

The least expensive type of flexible lining is known as a single skin flexible stainless steel liner. In this type of liner, the inner and outer surfaces are formed with the same piece of stainless steel, and so the inner surface of the liner is as corrugated as the outer surface. This is sometimes known as a class two liner. This liner should only be used with appliances that burn gas or oil, and must never be used with a solid fuel appliance.

Nearly all of these appliances are rated for flue gas temperatures of 250–260°C, but one or two are rated at 300°C. *Approved Document J* calls for a liner of 300°C for decorative fuel effect gas fires, so it is important to check the rating of the liner if a single skin flexible liner is to be used with one of these appliances.

For appliances with ratings above 300°C or solid fuel appliances, a double skin liner must be used. The inner and outer surfaces of these liners are formed with two different pieces of metal, and in many cases the grade of stainless steel used for each skin is different, as is the thickness. The outer layer provides the strength of the lining, while the inner layer, often only 0.2mm thick, creates a more acid-resistant and smoother inner surface. This product can be identified by the corrugated outer surface combined with the smoother inner surface.

As stated earlier, flexible stainless steel liners are often the quickest and easiest liners to install, but even then the installation is often carried out wrongly and so some information on this is given here. The lining itself must be in good condition and should be examined before the installation starts. In all cases there is a right and wrong direction for the liner to be put in the flue and nearly all types come with some kind of indicator to show which way the flue gases should travel in the lining. If the lining is installed the wrong way up this will not only reduce its life expectancy, but also invalidate any warranty offered by the manufacturer.

There are no British or European Standards in place for the installation of this type of lining, so the manufacturer's installation instructions must always be followed. The description below is based on the common features of such instructions, but the actual details must be checked before installation commences.

It is important to ensure that the flue is large enough for the flue liner to pass all the way down the chimney or the liner may become stuck or damaged during the installation. To do this, there are two methods usually quoted. One is to carry out a core ball test using a ball that is at least 25mm (1in) larger than the flue to be installed; if this passes down the flue under its own weight then the flue is large enough and the bends are gradual enough to allow the installation of the liner. However, the method favoured by manufacturers is to order an extra 1m length, and then cut this off and pass it down the flue from top to bottom. The test length should then be examined for damage or distortion before the full installation is carried out (if there is any damage to the test length the installation should not go ahead). If the test length is still in good condition, the installation can take place.

Initially, the chimney should be swept to ensure the removal of as much of the existing deposits as possible. It is recommended that sweeping is carried out from both the bottom and the top of the flue if possible.

Next, the chimney pot should be removed and the liner lowered down the flue. This should always be done from top to bottom (ensuring that the flue liner is the correct way up). A nose cone should always be used on the leading end to protect the flue liner and to lead the leading end through bends and offsets. The nose cone should have a rope attached that is lowered down the flue first, and then the cone connected to the liner using self-tapping screws. One person should be drawing the liner down the flue using the rope, while a second person is guiding the liner into the flue to avoid damage to the liner when it enters the flue.

Note that there should be no joints in the liner and the flue must be lined in one continuous piece. The liner must be wholly contained in the brick chimney and no part of the liner should be exposed. When the

liner has been passed down the flue it must be supported in place while any other works are carried out.

The liner then needs to be connected to the appliance. In the case of a single skin liner it is light enough to be supported by the clamp that will be placed at the top of the chimney later, but in the case of the double skin liners there will have to be a second support or clamp at the base. The appliance and offtake pipe should not support the liner and the appliance must be able to be removed without disturbing the liner.

An appliance or flue pipe connector is screwed to the base of the liner using stainless steel self-tapping screws and this is supported or secured using a clamp

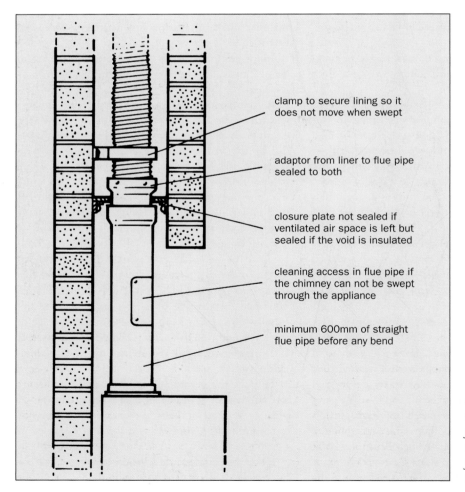

clamp to secure lining so it does not move when swept

adaptor from liner to flue pipe sealed to both

closure plate not sealed if ventilated air space is left but sealed if the void is insulated

cleaning access in flue pipe if the chimney can not be swept through the appliance

minimum 600mm of straight flue pipe before any bend

Fig 92 The connection of a flexible stainless steel liner to a flue pipe (top outlet).

Fig 93 The connection of a flexible stainless steel liner for an open fire.

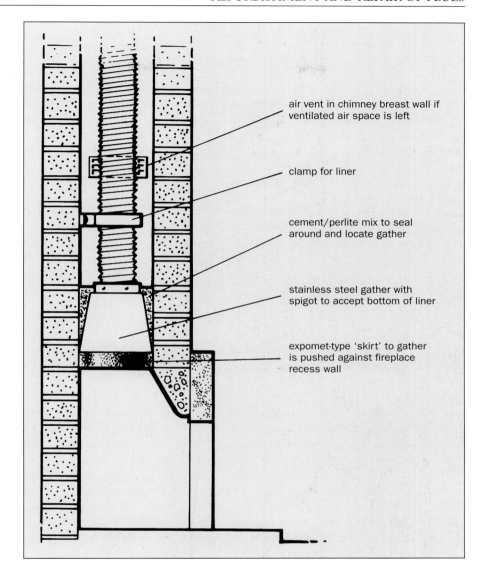

air vent in chimney breast wall if ventilated air space is left

clamp for liner

cement/perlite mix to seal around and locate gather

stainless steel gather with spigot to accept bottom of liner

expomet-type 'skirt' to gather is pushed against fireplace recess wall

and/or debris plate; in any case a debris plate must be fitted and fixed to the flue wall. If the flue is for solid fuel then the debris plate should be sealed so that any ventilation of the void is controlled or to prevent any insulation that is used escaping.

At this point it is important to decide whether the manufacturer requires the liner to be insulated or ventilated. If the liner is to be ventilated then the relevant vents must be fitted at the bottom and top of the flue. In some cases, this can be as much as 20sq cm at the base and two 15mm diameter vent pipes at the top. The bottom vent must be above the debris plate directly into the void, while the upper vent pipes should be in two different faces of the chimney stack at right angles to each other and placed into the chimney walls at an angle of at least 30 degrees up in to the stack to prevent water entering.

If the void between the liner and the chimney wall is to be insulated then this should now be added. No air vent at the base of the flue should be fitted, but

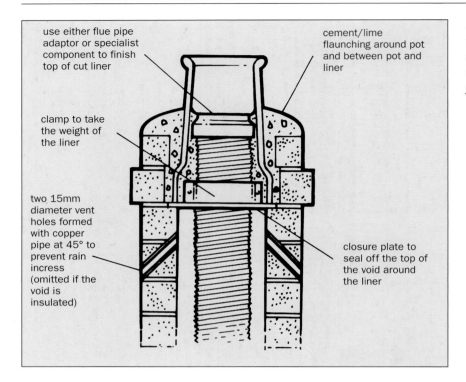

use either flue pipe adaptor or specialist component to finish top of cut liner

clamp to take the weight of the liner

two 15mm diameter vent holes formed with copper pipe at 45° to prevent rain incress (omitted if the void is insulated)

cement/lime flaunching around pot and between pot and liner

closure plate to seal off the top of the void around the liner

Fig 94 The location of vents at the chimney terminal, top plate and clamp.

some manufacturers require the upper vents to be fitted to allow the escape of any moisture contained in the insulation. Again, check with the manufacturer's instructions.

The choice of insulation will vary with flue liner manufacturers, but it is important that the correct type of insulation is used for the flue liner being fitted. If a loose insulation is to be used, this must completely fill the void and be dry and free from contamination or cement. The most common types used are vermiculite, perlite and leca (expanded clay beads). If the upper vents are to be provided at the top of the stack then it is not advised to use perlite as it is so light and fine that it may be blown out of the vents in windy weather.

The flue liner is then pulled tight at the top and a top plate and clamp installed. The top plate should be fixed to the top of the chimney stack and sealed to prevent rain ingress. There should be at least 75mm (3in) of flue liner protruding above the plate, but in the case of a solid fuel installation the liner should be long enough to pass at least halfway up the pot, and

then secured in place with a clamp. If the total length of the flue liner is more than 15m (49ft) an additional bracket or clamp should be fitted halfway up the flue.

The closure plate that can be used to support the mortar flaunching must make the top of the chimney stack watertight to prevent the ingress of water into the void or any insulation in that void. If this is not the case there is a risk of damp in the chimney breast further down the flue and an increased risk of corrosion of the liner from the outside.

If the liner is for a gas appliance the correct terminal should be riveted or secured with self-tapping screws to the top of the lining. In the case of a solid fuel appliance with the double skin liner, the liner should be cut to suit the terracotta chimney pot and then the inner and outer leaves either welded together or pop-riveted with stainless steel rivets. A top insert should then be fitted to stop water or condensation getting between the two skins of the liner and to prevent the sweep's brush damaging the end of the liner in the future. The chimney pot should then be placed over the projecting end of the liner and flaunched on using the normal mix for this task. The space between the chimney pot and the liner must be filled with the same mix used to flaunch the chimney pot and sloped to direct any water into the lining.

In all cases where a flexible stainless steel lining is used the chimney stack must be treated to make it waterproof, either by repointing and/or the use of a proprietary brick waterproofing compound.

Once complete, the lining must be tested in accordance with the Building Regulations for the fuel and lining types involved, a report completed and a notice plate fitted.

Coatings or Chimney Sealants

These are the most recent additions to the flue lining or chimney repair tool kit and they have already proved invaluable in the process of ensuring the safety of flues. The basic principle is to coat the inner walls of a flue with a suitable material and at the same time use a method of installation that forces the material into any gaps in the joints of the bricks or blocks to grout them and ensure a good flue seal. This type of system will leave only a thin layer of material on the flue wall and so is very useful where an open fire is connected to a flue of the correct size.

As the material is mixed on site it is imperative that the system used can demonstrate compliance with the Building Regulations section 220: (a) ii 'a cast in-situ flue relining system where the material and installation procedures are independently certified as suitable for use with solid fuel burning appliances.' If it is certified as above, it will also be suitable for gas and oil-fired appliances.

Advantages
• The area of the flue is changed very little when the lining is applied. Any of the other forms of lining will cause a major reduction in the area of the flue and so create an imbalance between the fireplace opening size and the flue cross-sectional area.
• There is limited disturbance to the decoration or the chimney breast unless there are long offsets or sharp angles within the flue.
• The lining can often be completed within a day.

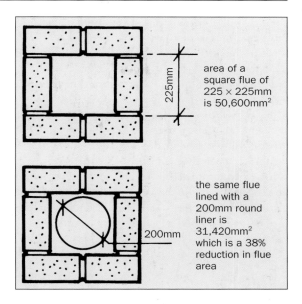

area of a square flue of 225 × 225mm is 50,600mm²

the same flue lined with a 200mm round liner is 31,420mm² which is a 38% reduction in flue area

Fig 95   The lining of a square flue with a round liner will reduce the flue area considerably.

Disadvantages
• Will only seal small holes in joints but not fill in missing brickwork.
• Will not reduce the area of the flue when installing stoves or smaller open fires.
• Will not form a new gather but can only coat the existing one.
• Must only be carried out by one of the manufacturer's certified companies.

There are a number of manufacturers of this type of system and all of the systems can only be installed by companies certified by the manufacturer. A full description of the lining process is therefore not required here, but again an understanding of the system will help in choosing a method as well as ensuring that the work is being carried out well.

With these systems, the most important part of the installation is the cleaning of the chimney. This is important in all types of flue lining but especially so in this case. The lining will be very thin and the material will not have enough strength to support itself, so it therefore relies on its adhesion to the chimney wall for strength. This means that any soot, dust, ash or tar that has not been removed will weaken the final lining. In addition to this, the cleaning process should be rigorous enough to remove any loose brick or mortar from the chimney wall.

To achieve this level of cleanliness, the normal sweeping of the chimney is not enough and the process will use steel-bladed brushes that are drawn up the chimney by a winch. Often there will be a 'train' of two or three brushes used and even then the process may have to be done twice to be sure of getting the walls clean and sound enough for the material to have good adhesion.

Once this has been done, the material is mixed on site in large enough quantities to complete the job if possible. The plunger will vary, but in the sequence shown here it is formed by the steel brushes with a sponge over them and then a hessian cloth placed over this. As can be seen, all three brushes are used, so that if any of the material falls past the first plunger it will fall onto the second and not be lost to the base of the flue or fall onto a bend and cause a partial blockage. The third plunger will again act as a final stopper for any material, but as it is not loaded with the material it will form the final smoothing pass to avoid a rough surface being left. As can be seen from the sequence of photographs here the

fireplace whoose flue is to be coated

flue before cleaning

health and safety always an issue

three wire brushes pulled through together

Fig 96 A series of photographs showing the lining of a flue using the Isokat chimney sealant method.

brushes being drawn through flue

flue clean and free of loose material

mixing the compound

setting up the plungers

drawing the plungers into the flue

the material is drawn up the flue by the plungers

the finished flue is sealed and smooth

completing the gather at the base

new pot fitted and roof cleared

completed flue from the bottom

Fig 96   (continued)

result is a smoother, completely sealed coating on the inside of the chimney. Where access is not easy at the gather of the fireplace it may be necessary to take out some of the chimney breast to cover this area completely and to ensure a smooth finish and complete coating. Note that the making good of the chimney breast will also require coating internally when complete.

Cast In-Situ Perlite Concrete

This is a well-established method of lining a chimney and has been in use for over twenty-five years. As with the above method, the material is mixed on site

and so the operator and the material must be certified as required in *Approved Document J* 2.20 (a). It is important to check that this certification is in place and that the operator either applies for building control approval or is certified for this type of operation by one of the competent persons' bodies. Since the materials are freely available and the process is not controlled by the manufacturers, it is important to ensure that the operatives are qualified and certified.

The system is basically to lower a rubber former down the flue, inflate it to the correct diameter and then centralize the resulting tube within the flue by spacing it away from the chimney walls. Once this

access holes are made at every bend and in any straight length greater than 3m

the tube can be seen at the access holes and spacers are added to make sure the tube does not touch the flue walls

the perlite cement is mixed using a specialist mixer and concrete pump

shuttering is added, along with wadding to support the concrete while it drys

Fig 97   A series of photographs showing the lining of a flue using the pumped perlite method of chimney lining.

Fig 97   (continued)

the perlite concrete is pumped into the flue through each access hole and the stones are replaced and held in place with shuttering boards

the final pumping is done from the top of the chimney stack until the void around the former tube is full

once the material has begun to set the soot door cover is removed and the cleaning access is re-made

when the shuttering is removed the stones are revealed and require pointing to match the wall

the tube is deflated, removed and the pot fitted

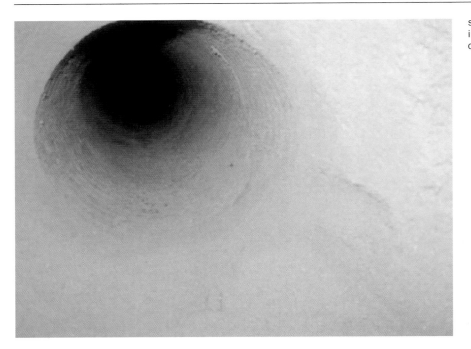

smooth, one piece, insulating flue when completed

Fig 97   (continued)

has been done, the perlite concrete is then pumped around the former in the flue and allowed to set. When the material has set, the former is deflated and removed.

Advantages

- The material is insulating and so keeps the flue warm.
- There are no joints except at the top and bottom of the flue, thereby reducing the possibility of leakage.
- The material has sufficient thickness to strengthen the brickwork within the flue and with special techniques it can be used to repair or replace the walls between two flues within a chimney.
- Has a life expectancy of over twenty-five years.
- Suitable for all fuels.

Disadvantages

- The centralizing of the former, especially at bends, requires holes to be made in the chimney breast at least every 3m (10ft). (At the time of writing there

is no fully certified system that does not require a hole in the chimney breast when there is a bend in the flue.)
- Will often take more than one day to complete.
- As a minimum wall thickness of 19mm (¾in) is required, the area of the flue will be reduced. (Sometimes, of course, this is an advantage.)

It is important that the lining engineer carefully surveys the chimney to ensure that there will be adequate room within the chimney to provide the minimum wall thickness of 19mm and the required flue diameter. If this is not possible, then this lining system should not be used. This survey must also be sufficient to identify that the chimney structure will be strong enough to support the amount of material to be pumped around the former. If this is not the case the operation must be carried out in steps to avoid too much strain on the structure, or a different type of lining should be used.

This survey must establish where the flue bends, in what direction and how far, as this will dictate where the chimney breast must be opened up to centralize the former in the chimney. Once this has been done, all the required access holes should be made and prepared.

When the above has been established, the former, which at this stage is a deflated reinforced rubber tube, is lowered down the flue from top to bottom and then inflated to the pressure needed to support the concrete to be pumped and to provide the correct diameter flue. Where the former passes the access holes, spacers are placed around the former to ensure that the minimum required material thickness will be in place at all points. The spacers used are of the same material as will be pumped around the former so that once set there will be no differential of expansion or problems of adhesion.

Once the former is in place the material is mixed in accordance to the certified code of practice and then pumped into the chimney. This pumping should be done through the access holes because:

- if the material is made wet enough to pump all the way from the top it may be weak;
- the material must not fall more than 3m (10ft) or flow more than 4m (13ft) or it may separate on impact or begin to set and prevent the correct flow of any further material;
- progress can be checked and the filling of the flue monitored.

Once one section is full, the chimney wall is made good with masonry and supported so that pumping into the next access hole can be started. This continues until the flue is cast, up to 100mm (4in) from the top of the masonry. The material is left to set for a period of time, which may vary from four to twenty-four hours depending on the material, temperature and the condition of the brickwork in the chimney. Once the material has set, the former is deflated and removed, leaving a smooth, one-piece, insulated flue liner.

Next, the chimney pot is put back in place. It must be bedded directly onto the top of the lining to ensure that there is no gap between the two. This bedding and flaunching is done using the mix described above in the section on bedding a chimney pot.

The flue should be tested as required by Building Regulations and then a bung placed in the bottom to avoid uneven drying of the material caused by the draw of dry air through the flue. The lining should be left like this for at least five days before any fire is lit, and even then the first fire should be small and limited to a short period.

Rigid Sectional Liners

These can be concrete, clay or stainless steel, but the principle is the same for each. However, it is important to realize that the stainless steel variety can only be used to reline an existing chimney, whereas the other materials are the same as those used when building a new chimney.

In this case, the lining is in sections of tube that can vary in length from 300–1,000mm (11¾–39in), and comes with bends that are produced in the factory. To get these liners into the flue will often require a considerable opening to be made in the chimney breast; the exception to this is when the flue is straight, as some types can be lowered down the flue from the chimney top.

All of these types of lining must have a socket and spigot joint and the socket must be facing up the flue when the liner is being installed.

Advantages
- The material is made in a factory under quality control and so will not vary.
- The liner is shaped and cut in the factory and so should provide a constant flue area.
- Some of the liners are made from an insulating material and are insulated around, making the flue warm.
- Some of these liners can have a life expectancy of up to sixty years.

Disadvantages
- There is a lot of disturbance to the chimney breast when this type of liner is installed in a bent flue.
- There are a large number of joints to be made, increasing the risk of leakage.
- The stainless steel varieties are only temporary, and although they will have a greater life expectancy than the flexible types they must still be removed when a new appliance is installed.

Fig 98   An angle iron support with starter block and flue pipe adaptor by Isokern.

- Some have thick walls and so only limited sizes can be installed into a standard brick chimney.

These materials and systems are available from most good builder's merchants and require little, if any, specialist equipment to install. However, it is still important to use a professional chimney engineer and ensure that they are either registered as competent persons for this work or that building control permission is obtained.

The process is a little like laying a drain in the ground, but on a vertical plane. Each liner must be correctly jointed to the one before it, fully supported in all directions and covered over. All the joints must be as clean and smooth as possible and no material must be left to create a partial blockage.

The installation starts with creating a good support for the liners. The form this takes will depend on a number of things, but the most important is whether the flue is for an open fire or a closed appliance such as a stove or boiler.

If an open fire is to be used, then a gather must be formed. Most manufacturers will provide the components needed for this or a gather can be formed in brickwork, remembering that in *Approved Document J* it states that any gather must be smooth.

If a closed appliance or stove is to be installed the liner could be supported on angle iron or concrete lintels, a starter block and flue pipe adaptor.

From this point it is important to ensure that the liners are the correct way up and that the first liner is sealed well to the gather or starter block. Each joint must be made with both the jointing compound and any secondary strapping system required by the manufacturer.

When each joint is made, the liner should be back-filled with insulating cement. This should be lightly tamped down to ensure there are no gaps in the insulation and to ensure that the liners are securely supported by it.

The process continues through the flue, with extra care taken at the bends and the offset – at these

chimney cap removed and first liner placed into the top of the flue

liner supported on a rope so it can be lowered down as the lining is assembled

next liner is added using jointing compound and a wide band to support the joint while lowered down the flue

this is then continued as the flue is constructed and passed down the chimney

rods are used to bring the liner to meet the starter block

the lining is jointed to the starter block and the frame is filled in to support the insulating backfill

Fig 99 A series of photographs showing the lining of a straight flue using Isokern liners *(continued on page 120).*

points, ensuring that the liners and bends all line up and are correctly jointed and sealed is very difficult, and may well require almost the entire front of the chimney breast to be removed in this section.

The liner must carry on through to the top of the chimney stack and then be sealed to the chimney pot, which must be flaunched and supported as described in the section above on replacing a chimney pot.

There are many possible problems and variations on this basic description, as many of the product manufacturers specify different installation methods and chimneys vary considerably.

In the illustration here it can be seen how a lining can be installed into a straight flue. This method can be used in different stages where there are long, straight sections before and after an offset.

The connection of the liner to the appliance should be carried out by a competent person such as a CORGI, OFTEC or HETAS-approved installer, or there must be building control permission in place for the installation (if a gas appliance is connected to the gas supply and a new liner is to be connected to the appliance, this must be carried out by a CORGI-registered installer).

the leca backfill is mixed and poured around the lining

a bag is dragged through the flue to clean all the joints

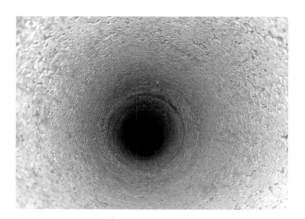

the flue is smooth with clean surfaces and no protruding jointing material

pot is added and lining flaunched over to seal top of chimney

CHAPTER 9

Appliance Installations

As it is not legal for a non-CORGI-registered installer to fit a gas appliance, only very limited information on this has been included in this book, in order to enable you to identify the types of appliances and some of the variations in installation practice. In addition, there is also some information on the installation of oil appliances where this practice will vary from that for a solid fuel or wood-burning appliance of the same type.

OPEN FIRES

It is often considered that there is no actual installation to a solid fuel open fire – you just have to put a grate in the opening and light the fire. In some instances, if all the previous operations have been carried out correctly, this is the case, but that is certainly not always so. Indeed, if an open fire is to be efficient and effective then there will be some installation involved whether it is for a gas appliance or a solid fuel one.

There is a wide variety of open fires, in appearance, performance and installation methods. We will start with the installation of a solid fuel open fire that conforms to the British Standards, even though this type of fire installation is becoming less popular. If the installation is carried out like this it will also be suitable for a gas decorative fuel effect (DFE) fire, but a number of the gas DFE fires do not need this full installation as some can be fitted into a box or chamber that comes pre-insulated.

INSET OPEN FIRES

It is important that any fire surround will fit directly against the brickwork of the chimney breast and so any plaster in the area to be covered by the fire surround must be removed. The back hearth should be correctly positioned and levelled. It is important to ensure that the superimposed hearth (which will be fitted later) and the back hearth will end up at the same level.

The fireback must be a two-piece fireback manufactured in accordance with BS 1251; if a one-piece fireback is purchased it must be cut along the score line to make the correct two pieces. The fireback should be correctly positioned on the back hearth. This can be confirmed by making a trial fitting of any fire surround, making sure that the opening in the surround matches the front edges of the fireback and that the fireback is plumb and set 12mm (½in) back from the chimney breast. This position should be marked to ensure the fireback does not move during the remainder of the installation. The trial fireplace is then removed and the installation is continued as stated below.

The top half is then placed on the bottom half; neither section is mortared down to the hearth or each other. Next, the space at the side of the fireback is bricked up, but the sides of the bricking up in contact with the fireback are not mortared.

Corrugated cardboard is placed behind the lower half of the fireback and the space behind is infilled with an insulating mix of six parts vermiculite to one part OPC and just enough water to wet the mix. The mix is tamped down very lightly until level with the top of the fireback.

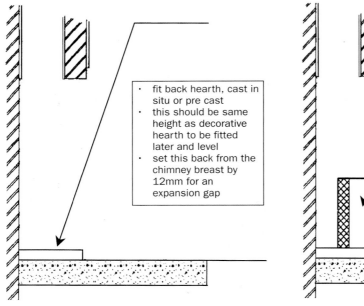

- fit back hearth, cast in situ or pre cast
- this should be same height as decorative hearth to be fitted later and level
- set this back from the chimney breast by 12mm for an expansion gap

1. clear out all the old rubble and check the constructional hearth, add in a new back hearth setting it 12mm back from the chimney breast

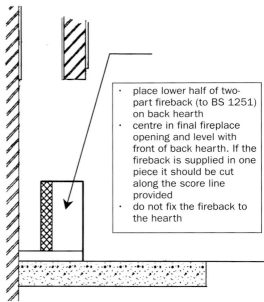

- place lower half of two-part fireback (to BS 1251) on back hearth
- centre in final fireplace opening and level with front of back hearth. If the fireback is supplied in one piece it should be cut along the score line provided
- do not fix the fireback to the hearth

2. place and centralize lower half of fireback onto the back hearth, set the front level with front of back hearth

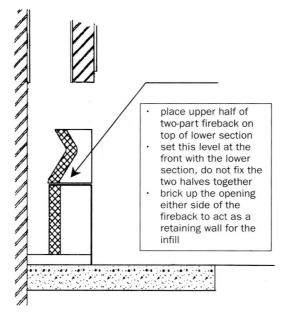

- place upper half of two-part fireback on top of lower section
- set this level at the front with the lower section, do not fix the two halves together
- brick up the opening either side of the fireback to act as a retaining wall for the infill

3. place top half of the fireback onto bottom half and build up the sides with brickwork to act as shuttering

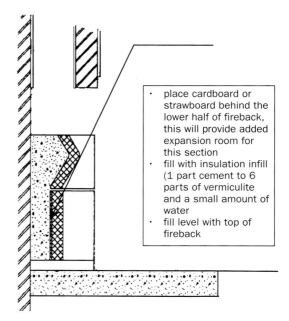

- place cardboard or strawboard behind the lower half of fireback, this will provide added expansion room for this section
- fill with insulation infill (1 part cement to 6 parts of vermiculite and a small amount of water
- fill level with top of fireback

4. put corrigated cardboard behind lower half of fireback and then infill behind with vermiculite cement

Fig 100   This series of diagrams shows the installation of a British Standard fireback and fireplace.

122

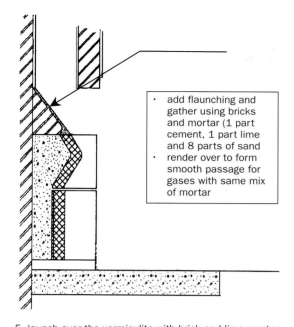

- add flaunching and gather using bricks and mortar (1 part cement, 1 part lime and 8 parts of sand
- render over to form smooth passage for gases with same mix of mortar

5. launch over the vermiculite with brick and lime mortar

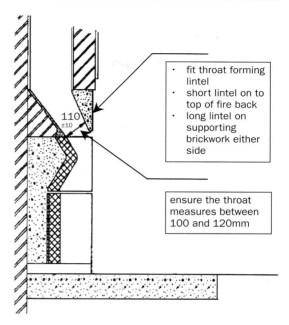

110 ±10

- fit throat forming lintel
- short lintel on to top of fire back
- long lintel on supporting brickwork either side

ensure the throat measures between 100 and 120mm

6. position throat forming lintel to provide a throat depth of 110mm±10

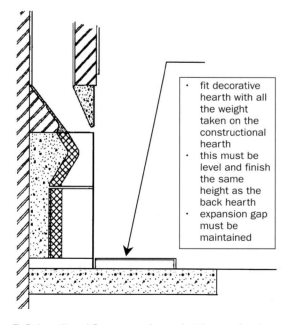

- fit decorative hearth with all the weight taken on the constructional hearth
- this must be level and finish the same height as the back hearth
- expansion gap must be maintained

7. fit hearth and fire surround ensuring they are level

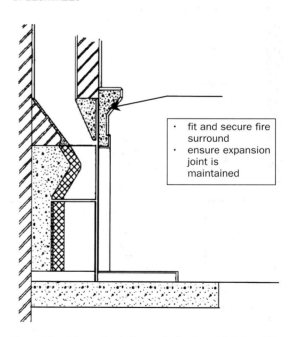

- fit and secure fire surround
- ensure expansion joint is maintained

8. seal the 12mm gap between fireback and lintel and the hearth and surround

Fig 100   (continued and overleaf)

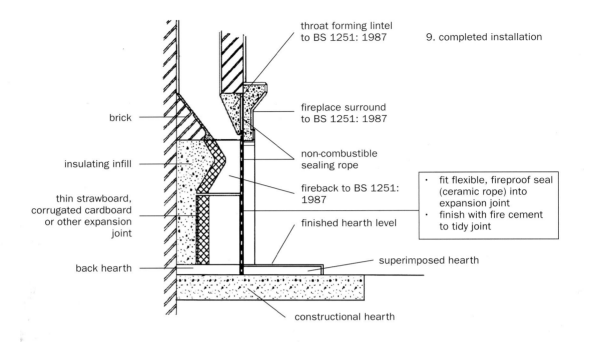

From this point, the flaunching from the back of the fireback is to go up at an angle of 60 degrees from the horizontal to the back of the gather and at the sides the flaunching goes up vertically until it meets with the sloping sides of the gather unit. This flaunching can be made of brickwork but must be rendered over using a mix of one part OPC, one part lime and eight parts of sand.

The gather unit insert that forms the throat-forming lintel should now be bedded in place using the refractory mortar previously used. The superimposed hearth, in accordance with BS 1251, should be laid to meet *Approved Document J* of the Building Regulations and BS 8303.

The hearth must be at least 48mm (1⅞in) thick plus a decorative surface and should extend 400mm (15½in) in front of the chimney breast. The hearth must extend at least 300mm in front of the final fireplace opening and be at least 150mm (6in) either side of the finished fireplace opening.

The fireplace surround should be no more than 65mm (2½in) thick for a distance of not less than 50mm (2in) either side, and above, the finished fireplace opening; the fire surround must extend at least 1.2m (47in) above the constructional hearth level if it is to be topped by a wooden mantle shelf.

The final opening size of the fireplace opening should result in the bottom of the throat lintel being between 5–25mm (⅕-1in) above the top of the opening.

The 12mm (½in) gap between the fireback, the fire surround and hearth is to be caulked with a mineral rope and skimmed with a thin skim of fire cement.

A more popular type of appliance these days is to have a cast iron insert in front of the fireback and in this case the installation is the same. However, with the wish for more and more accuracy in the appearance of the finished fireplace there are more fireplaces available that are reproductions and include a steel or cast fireback as part of the insert. This type of installation does away with the British Standard components, and as a result will often be less efficient and cause a greater draught in the room.

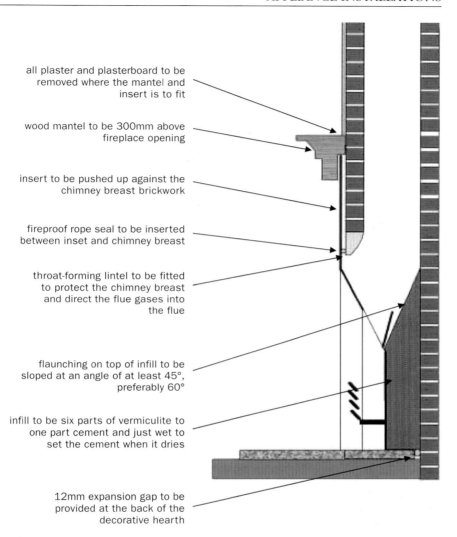

Fig 101   The correct installation of a cast iron inset or register grate.

all plaster and plasterboard to be removed where the mantel and insert is to fit

wood mantel to be 300mm above fireplace opening

insert to be pushed up against the chimney breast brickwork

fireproof rope seal to be inserted between inset and chimney breast

throat-forming lintel to be fitted to protect the chimney breast and direct the flue gases into the flue

flaunching on top of infill to be sloped at an angle of at least 45°, preferably 60°

infill to be six parts of vermiculite to one part cement and just wet to set the cement when it dries

12mm expansion gap to be provided at the back of the decorative hearth

These installations must still be carried out with the same principles being adhered to as for the previous types: keep the air entering the flue to a minimum; keep the air moving as fast as possible; and avoid large spaces and sudden changes in area. If this is done and the chimney has been correctly assessed and repaired if required, there should be few, if any, problems.

The illustration here shows a cast iron insert installed correctly, for both gas and solid fuel. As can be seen, the filling behind the fireback part of the insert is with an insulating material. The top of this is then flaunched in the same way so that the gather space is kept to a minimum and there is no area for the build-up of soot or deposits. This flaunching also provides a smooth route for the products of combustion and this should be matched as near as possible by the fitting of a throat-forming lintel just behind the front section of the insert and as close to the top of the opening as possible. The gap between the insert and the chimney breast must be sealed with a flexible seal (the metal insert will expand and contract more than the firebrick will), such as fireproof rope and so on.

If all of these issues are considered, the operation of the fire, irrespective of the fuel in use, will take all the products of combustion up the flue at all times.

Large Fireplace Design

With all successful open fires the reason they work without any smoke entering the room is that the air is drawn into the fireplace and up the flue fast enough to drag any smoke or fumes being produced back into the gather and up the chimney.

To achieve this, the speed of the air entering the fireplace must be an average across the whole of the opening of at least 0.3m per second (60ft per minute). This means that the bigger the fireplace opening, the more air must be drawn into the opening and gather to achieve this speed. The more air drawn into the fireplace and gather means the bigger the flue in the chimney must be to be able to transport it out to the atmosphere.

The larger the fireplace, the greater the amount of air needed to ensure that the required speed is reached and to avoid spillage of fumes or smoke into the room. As stated earlier in the principles of how a chimney works, the taller the flue the greater the draw and so the more energy it has. This means that a tall chimney can make the air move faster in the flue and so will be able to make a bigger fire opening work efficiently.

When designing a fireplace larger than is standard the skill is to bring together the correct flue diameter, chimney height and fireplace size. There are many other factors that will affect the performance of a fireplace, but to consider these as well is beyond the scope of this book.

As related in Chapter 6, the Solid Fuel Association produce a graph to help with this. **It is important to be aware that if the graph results in a flue size of greater then 300mm (11¾in) diameter or 0.07m^2 (0.75ft^2) flue area then expert advice must be sought as the graph cannot take into account all the factors that have to be considered.**

Also in Chapter 6 is a series of diagrams to show how to calculate the area of a fireplace opening in the more common configurations, including the fitting of a canopy as discussed later.

In the case of any open fire without a throat, the free area of the air vent into the room must equal at least half the area of the flue serving the fireplace, and so the smaller the flue that can be made to operate, the smaller the air vent that is needed.

It is also important that the gather from the top of the fireplace to the base of the flue lining is smooth and at no point forms an angle greater than 45 degrees from the vertical.

Calculations for Three Popular Fireplace Sizes

To give a better indication of the air movement caused by different-sized open fires here are some examples. We will look at three different fireplace sizes that are popular with designers and architects:

- Firstly, let us look at the British Standard fireplace discussed above:
  550mm high × 400mm wide gives a fireplace area of 0.22m^2.
  At a speed of 0.3m/s the volume of air taken into the chimney every hour will be 237m^3.
- In the second example we will take a basket grate in a larger opening:
  600mm high × 600mm wide gives a fireplace area of 0.36m^2.

At a speed of 0.3m/s the volume of air taken into the chimney every hour will be 388m^3.
- In the third example we will take the very popular large fireplace opening of 1,000 × 1,000mm, which is very common in larger rooms:
  1,000mm high × 1,000mm wide gives a fireplace area of 1.0m^2.
  At a speed of 0.3m/s the volume of air taken into the chimney every hour will be 1,080m^3.

It can be seen from this that once the fireplace becomes larger the increase in air movement through the room increases dramatically. The flue size will therefore have to increase considerably, as will the air vent, to allow this amount of air to enter the room.

Fig 102   A dog basket in a large opening with a canopy fitted.

Canopies over Basket Grates

As has been stated earlier, the use of a large open fire is very inefficient, but they nonetheless hold considerable aesthetic appeal. It is possible to fit a basket grate under a canopy as shown here, or to fit a free-standing open fire into the opening. Both of these alternatives will retain the large opening appeal while increasing the efficiency of the fire and reducing the air movement.

The fitting of a canopy is not difficult, but as with the large fireplace the design is very important and is crucial to the success of the fireplace. With gas, the Gas Safety in Use Regulations require the bottom of a canopy to extend beyond the edges of the fire by a considerable amount and in most cases this makes the use of a canopy over a gas fire impractical. The

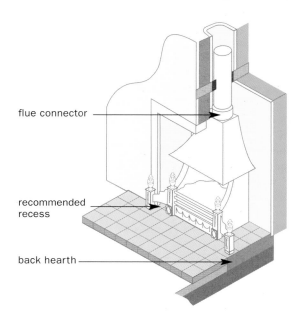

flue connector

recommended recess

back hearth

Fig 103   A free-standing open fire fitted in a large fireplace opening.

overlap makes the base of perimeter of the canopy so large that the effective opening size is far too great to operate with a standard size flue liner. If it is intended to use a canopy over a DFE gas fire, consult a specialist in the design of these arrangements as problems can occur very easily.

In the case of a solid fuel or wood-burning grate there are no such regulations and so the design can be simplified. Remember when designing a canopy that the intention is to reduce the effective area of the fireplace opening, reduce the amount of air entering the chimney and so increase the efficiency or reduce the flue size needed. This is important, as a badly designed canopy will result in a larger effective fireplace opening area and so give greater problems than using the fire without it.

The illustration on p.68 shows the way in which the effective fireplace opening area for a canopy is calculated. The first rule is that to keep the effective fireplace opening as small as possible the edge of the base of the canopy should be as near in size as possible to the size of the grate that will fit under it. The maximum recommended overlap is 50mm (2in). The height above the hearth should be reduced as much as possible without causing problems of obscuring the view of the fire. If these simple rules are followed, a canopy should provide a reduction in the fireplace's effective opening area.

Also important in the design of a canopy is the smoothness of the gas flow once the smoke and fumes have been drawn into it. The sides of the canopy should start with a vertical section and then slope over to meet the flue. This slope should be 30 degrees from the vertical if possible, but never more than 45 degrees from the vertical. Just before entering the flue the canopy sides should return to the vertical and the top of the canopy should match the size and shape of the liner in the flue as near as possible.

The illustration on p.127 shows a well-shaped canopy that has been produced by Chandlers, which is a company that specializes in canopy design and installation. The very smooth nature of the change of angles and shape can clearly be seen, as can the closeness of the size of canopy base and grate.

Once designed, the canopy has to be installed correctly to ensure good operation. The canopy must be fixed in place and sealed to the walls of the chimney or the flue liner, and the back of the canopy must be sealed to the back wall of the fireplace opening.

It must be understood that canopies should be designed and made specifically for the fireplace into which they are going to be fitted. This is especially the case where they are being installed into an existing or old fireplace and flue.

Inset Open Fire Convector Units

This is a development on the standard open fire, and has a built-in convection chamber to enable cool air from the room to be drawn around the back of the fire to collect more heat and so operate more efficiently. There are a number of these types of fire – some are basic boxes, while others are more complicated and effective.

The installation of this type of appliance is similar to that of an inset stove or room heater, in that they are self-contained units that are sealed to the chimney via a flue pipe or gather. One type of inset convector is the firebox, which is typified by the Jetmaster unit. These come in many sizes and normally contain an adjustable throat damper. The convection chamber is formed as a part of the appliance; the air at current room temperature is drawn in at hearth level and passed through a duct under, behind and over the fire, before being released as hot air into the room just above the fireplace opening.

The unit is installed into a builder's opening and can either make use of the chimney gather or be fitted with a made-to-measure gather. As the shape and size of the flue outlets can be unusual, it is often better to use the factory-made gather unit, which will have an outlet formed on it to enable a flue pipe to be fitted and sealed into the flue liner. This makes a positive connection between the appliance and the liner, thus ensuring efficient gas movement and avoiding leaks.

The second type of appliance is those that do not have a built-in convection chamber but rely on the installer creating this as part of the installation process. This type of installation is more complicated, but can have a number of advantages. The air can be drawn into the convection chamber directly from outside (this will form all or part of the ventilation air required under Building Regulations), and is

Fig 104   A convector firebox fitted into a fireplace.

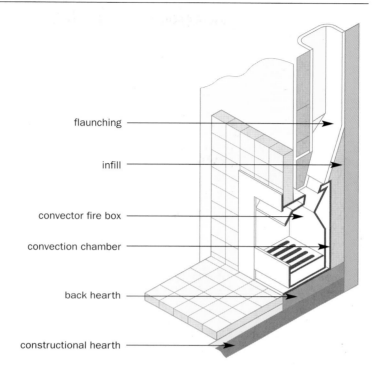

flaunching

infill

convector fire box

convection chamber

back hearth

constructional hearth

then drawn up around the body of the appliance, which has heat-sink type fins on it, and is then taken past the uninsulated flue pipe within the chamber, before being expelled out of the chamber through outlets either in the same room as the appliance or another room. The hot air from the convection chamber can be ducted around the house using fans, enabling this single appliance to warm several parts of the property.

Shown overleaf is a typical installation of a Dovré 2000 series inset convector open fire. This appliance has foldaway doors that can be closed, which includes the construction of the convection chamber. This type of installation is special to each appliance and should be carried out strictly in accordance with the manufacturer's installation instructions.

Free-Standing Convector Open Fire

These are the most commonly used alternatives to basket grates in large or inglenook fireplaces. The reason for this is that these appliances can be fitted into a large fireplace opening and provide an open fire that has a greater heat output (due to the larger surface area), more convected heat and higher efficiencies than the basket grate. Because they have factory-formed throats and canopies they can also provide a larger view of a fire while operating safely and effectively on the smaller 200–225mm (8–9¾in) standard flue liner.

The installation is the same as for the free-standing closed stoves and boilers inside a fireplace recess, and so to avoid repetition the details of this will be discussed in that section below. As always, it is important to ensure the suitability of the chimney and for the appliance to be installed strictly in accordance with the manufacturer's instructions. Examples of this type of appliance include the Rayburn Rembrandt shown on p.127.

STOVES AND ROOM HEATERS

As well as there being a number of different types of stoves and room heaters there are also various methods of installation, several of which are covered here.

Fig 105 A
Dovré 2000 inset
with a convector
chamber built
into the chimney
breast.

Within a Fireplace Recess

If the stove is a solid fuel appliance in a fireplace recess, irrespective of the appliance type it must have a constructional hearth that fills the recess and extends at least 500mm (19½in) in front of the chimney breast and 150mm (6in) either side of the recess. This constructional hearth must be at least 125mm (5in) thick and be of solid non-combustible material.

Inset Room Heater with Surround Seal

This type of appliance is the easiest and cheapest way to change an open fire to a closed appliance and gain the advantages of higher efficiency and better control. Some models can be used with the original open fire components still in place, hence their being known as 'fire converters'. Some of these appliances use the space between the back of the appliance and the original fireback as a convection chamber, while others have a double-wall convection chamber built in. If a type that uses the fireback as part of the convection chamber is to be installed, it is vitally important to get this as clean as possible to avoid the smell of soot being drawn out into the room with the convected heat.

After the chimney has been swept and checked as normal, any grate or other components must be removed from the opening to make it free and clean from obstructions. The new appliance is placed in the opening and centred, the securing hole is marked and then the appliance removed. The securing hole is drilled and the appropriate fixing is placed in the hole. The appliance is then put back in the opening, but with a flexible fireproof rope adhered to the back of the unit to be trapped between the appliance and the fire surround. The base is screwed down to the hearth via the fixing and is pulled in at the top using the clamp mechanism supplied. This clamp brings the appliance right up against the seal and surround to make a tight fit.

If the unit has a built-in convection chamber then the space between the back of the appliance and the fireback is filled with insulating cement up to the level of the outlet.

It is also possible to buy a door set that can be secured to the fire surround, and therefore still use the fireback as the back of the appliance. These doors can come as simple doors as shown overleaf, or as a unit that includes controls and a convection chamber.

Inset Room Heater with Chimney Seal

In the past, this has been the most popular type of installation, since if a chimney breast was present it used little, if any, extra space in the room.

There are still many of this type of installation being fitted, with many appliances being imported from abroad that use this method, although the details may vary.

- In the most basic form any existing fireplace is removed and the chimney breast opened up to reveal the builder's opening and the original gather.
- The appliance is placed in the opening and centralized to ensure the flue can be connected to the appliance without having to use tight bends.
- A back hearth is cast or placed on the base of the builder's opening to bring the surface up to what will be the final height of the finished decorative hearth.
- An access hole is made in the front of the chimney breast above the builder's opening to allow the fitting of the offtake pipe and the required insulation.

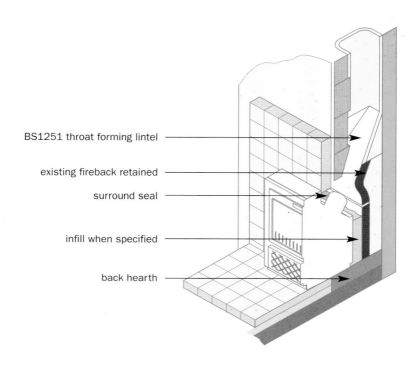

BS1251 throat forming lintel

existing fireback retained

surround seal

infill when specified

back hearth

Fig 106   Inset stove with surround seal, converting an open fire to a closed fireplace.

131

Fig 107   A glass door set used with an open fire to reduce air movement in the room.

- The fire surround is put in place, centralized on the chimney breast and secured in position.
- The appliance is pushed through the fireplace opening, pushed hard against the surround with a flexible rope seal between the two, sited on this back hearth, centralized, then secured in place.
- The offtake pipe is placed into the outlet spigot of the appliance and sealed in place using a rope seal and fire cement. In this situation, where the offtake pipe will be insulated around, only 5mm thick cast iron to BS 41, 3mm thick mild steel flue pipe complying with BS 1449 Part 1 or 1mm

thick stainless steel to BS EN 10088–1 can be used. This pipe should be long enough to bridge the gather if one exists and pass into the flue. The pipe is sealed into the flue liner if the size is correct. If the offtake pipe diameter is less than the flue size it should finish at the top of the gather.

- The space around the appliance and the offtake pipe is filled with an insulating cement of six parts vermiculite, one part OPC and a small amount of water to enable the mix to set. This is taken up to a level of 25mm (1in) below the top of the offtake pipe. Above this there is flaunching made from

Fig 108 A typical installation of a solid fuel inset room heater with a chimney seal.

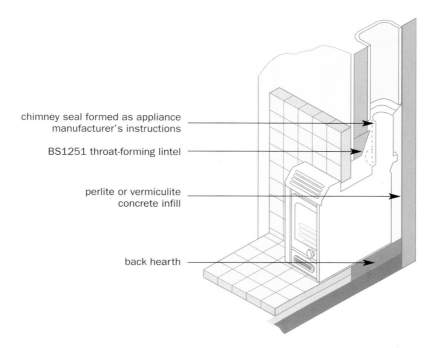

chimney seal formed as appliance manufacturer's instructions

BS1251 throat-forming lintel

perlite or vermiculite concrete infill

back hearth

one part lime, one part OPC to eight parts of sand and sufficient water to make a mortar mix, fitted to finish at an angle of 45 degrees from the vertical, so directing any deposits back into the offtake pipe and sealing the offtake pipe to the walls of the flue.

Many of the appliances coming in from abroad have convection chambers, and in some cases even forced, ducted warm air systems that can warm more than one room when the fire is lit. This type of installation is often considered easier and cleaner, and to increase the ease of installation still further it has an offtake pipe as part of the appliance. This can be fitted to the appliance by dropping the offtake spigot into the firebox so that a flexible flue lining can be connected without the need for the access hole in the chimney breast.

Free-Standing Room Heater in a Fireplace Recess

The fireplace recess used here should provide adequate space for the appliance and any maintenance that it will require. The appliance manufacturer will often quote a minimum space around the appliance and this must be adhered to. Great care must be taken when considering the access for chimney sweeping as it must be possible to sweep the chimney clean with an appropriate size and grade of sweep's brush and rod to suit the type of fuel and the size of flue.

To avoid creating a fire risk when installing an appliance with an exposed flue pipe, as in these installations, adequate clearance from combustible materials must be ensured. This is important because not only will the pipe get hot under normal operation (up to 400–450°C), but in the event of a chimney fire the flue pipe can reach temperatures in the region of 1,000°C (indeed, I have seen a flue pipe glow red in these conditions). Because this can occur at any time, combustible material must be kept well away from the flue pipe, or be insulated or be shielded.

There are two types of installation that can take place in a fireplace recess and which one is chosen will depend on a number of factors, but the most common factor is whether the stove comes with a top or back outlet.

Clearances Between a Flue Pipe and Combustible Material

The greatest clearance from a flue pipe to combustible material is that needed for a solid fuel stove. In this case, if the combustible material is unshielded and the flue pipe uninsulated, then the clearance should be three times the outside diameter of the flue pipe. For a 150mm (6in) flue pipe this means a clearance of 450mm (18in).

This is often difficult to obtain, so a shield of non-combustible board can be placed in front of the combustible material with an air gap of 12mm (½in) between them. If this is done, the clearance can be reduced to one and a half times the outside diameter of the flue pipe. For a 150mm flue pipe this means a clearance of 225mm (9in).

The Building Regulations also offer the option of insulating the flue pipe with 12mm (½in) of insulation with a specific insulating value, or the use of a prefabricated chimney system directly off the appliance. In the latter case, this can often get as close as 50mm (2in) to wood or other combustibles.

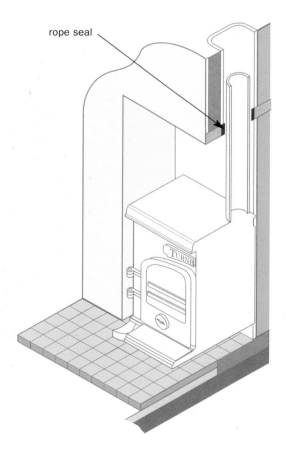

rope seal

Fig 109 A free-standing stove with a top flue outlet fitted in a fireplace recess with a lined flue.

Top Outlet Stoves

In this type of appliance the outlet spigot is fixed on the top and the flue pipe goes straight up into the flue. According to the relevant British Standard the flue pipe should be straight and vertical for 600mm (23in) and should be of cast iron, mild steel or stainless steel as in the inset method, but with the additional option of being able to use vitreous enamelled steel pipe complying with BS 6999.

The illustration here shows this type of appliance installed into a builder's opening that has been constructed with a flat raft lintel with a flue liner built on the top. This is the ideal situation for this type of installation when a new fireplace or chimney is being constructed. If the flue is not already lined and the appliance has a high efficiency, consideration should be given to the need to line the flue to bring the size and volume down to suit the appliance.

If the flue is to be lined with a flexible stainless steel liner the flue pipe should be sealed directly to the liner and the top of the opening should be closed off with a non-combustible closure plate.

If the flue is not to be lined then the installation will be very different and great care must be taken to avoid leaving a large space above the offtake pipe, as this can affect the performance of the appliance and be dangerous.

The flue pipe should pass up through the gather and finish in a position where the width of the gather and the register plate to be fitted is no more than three times the external diameter of the flue pipe. Any greater than this will result in a void that will cool and slow down the gases dramatically, the effect

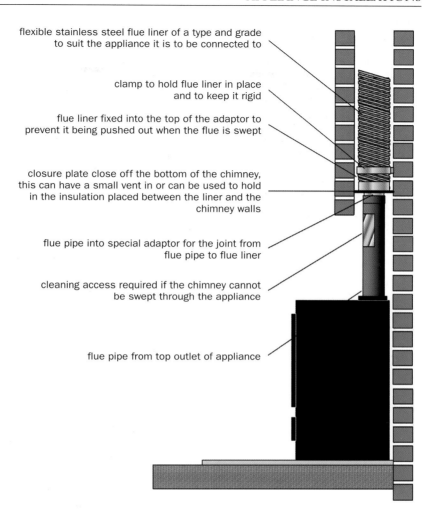

Fig 110 A free-standing stove with a top flue outlet fitted in a fireplace recess and connected to a flexible stainless steel flue liner.

flexible stainless steel flue liner of a type and grade to suit the appliance it is to be connected to

clamp to hold flue liner in place and to keep it rigid

flue liner fixed into the top of the adaptor to prevent it being pushed out when the flue is swept

closure plate close off the bottom of the chimney, this can have a small vent in or can be used to hold in the insulation placed between the liner and the chimney walls

flue pipe into special adaptor for the joint from flue pipe to flue liner

cleaning access required if the chimney cannot be swept through the appliance

flue pipe from top outlet of appliance

of which will be to reduce the buoyancy of the gases and increase the sooty or tarry deposits formed at this point.

British Standards state that the register plate can only be of 1.5mm-thick rust-protected mild steel or 100mm (4in) of solid non-combustible material such as concrete. If steel is chosen, the sealing of the register plate to both the walls of the flue and the flue pipe is very important. Where the fuel being burnt is a smokeless fuel or bituminous coal the amount of moisture in the fuel is relatively low and so on lighting the appliance in a cold flue the level of condensation is small and can therefore be absorbed by a fireproof rope. For this reason, when a flue pipe passes through a concrete lintel or into clay or concrete liners this is the material used to form the seal. In the case of the concrete lintel the rope is wrapped around the flue pipe three or four times and then packed into the space between the flue pipe and the lintel. This method provides a continuous run of seal with no joint to form a weak point and will provide enough rope to absorb the condensates formed.

This rope seal is then held in place using a clamping ring around the flue pipe. This will tidy up the joint as well as ensure that the rope stays in place. When the fire is lit, the flue pipe will expand and increase its diameter so as to squeeze the rope and

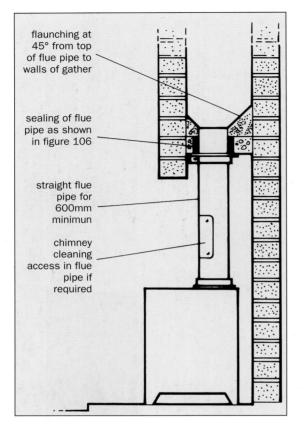

Fig 111   A free-standing stove with a top flue outlet fitted in a fireplace recess with a raft lintel.

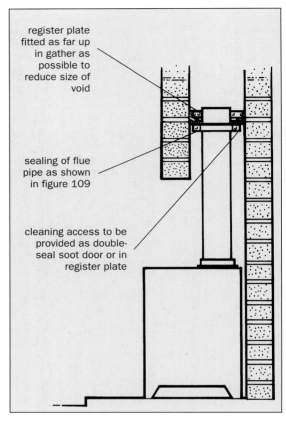

Fig 112   A free-standing stove with a top flue outlet fitted in a fireplace recess with a register plate.

form a tighter seal. If a fire cement is used as a seal in this place, the expansion and contraction of the flue pipe, when the fire is lit and allowed to go out, will result in the cement cracking and falling out over time, which will break the seal.

In the case of the steel register plate there are more seals to deal with and so the installation must be carefully carried out. First the supporting frame has to be put in place and sealed to the walls of the gather. Once this is done the register plate has to be secured up against the frame with a fireproof-rope seal trapped between the two. The register plate should be secured to the underside of the frame so that as the flue pipe expands in length the force of the expansion squeezes the rope and adds to the seal rather than trying to split it open. The flue pipe is passed through

the hole in the register plate, but because of the thin edge of the plate it is not possible to pack the joint with fireproof rope or fire cement. For this reason, the fireproof rope is run round the top flange of the clamping ring, and then this is pushed up against the register plate and trapped as the ring is clamped to the flue pipe.

The installation of an appliance in which it is intended to burn wood will require a slightly different approach, the reason for this being the higher moisture content of the fuel. Even when wood has been dried and correctly prepared for burning it will have up to three times the moisture content of a smokeless fuel and so the amount of condensation that will form on lighting the fire will be considerably more. As a result, it is not wise to rely on a rope seal

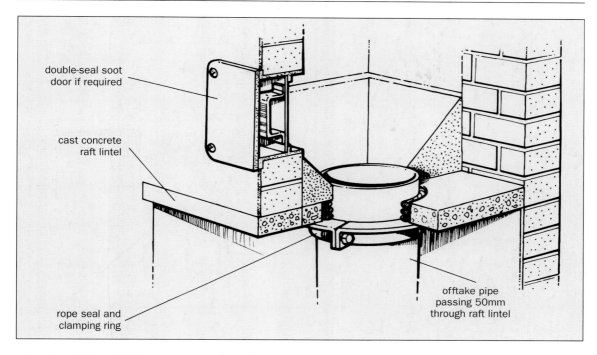

double-seal soot
door if required

cast concrete
raft lintel

rope seal and
clamping ring

offtake pipe
passing 50mm
through raft lintel

Fig 113   Sealing of a flue pipe into a raft lintel.

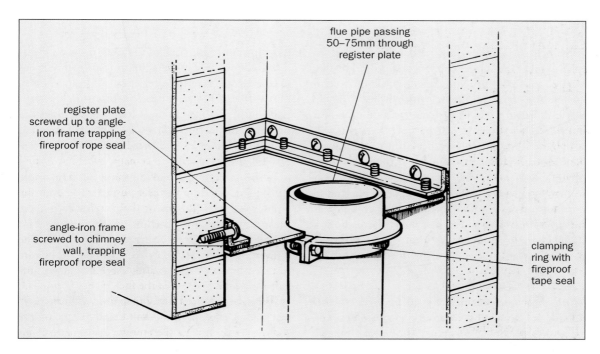

flue pipe passing
50–75mm through
register plate

register plate
screwed up to angle-
iron frame trapping
fireproof rope seal

angle-iron frame
screwed to chimney
wall, trapping
fireproof rope seal

clamping
ring with
fireproof
tape seal

Fig 114   Sealing a register plate and flue pipe.

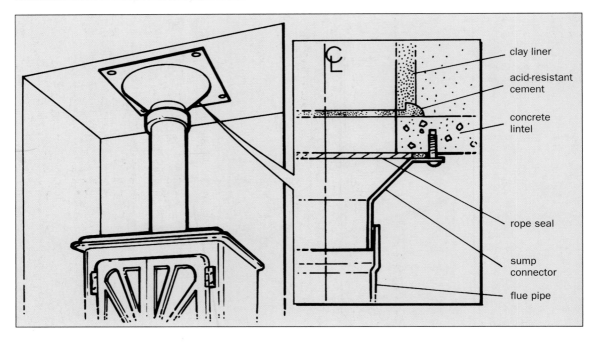

Fig 115   Using a sump connector with a concrete lintel and a wood-burning stove and a lined flue.

to absorb the condensation and so a method of ensuring that the condensates are retained in the flue is required. To do this, most installers will fit an item known as a sump connector, which is like a funnel that is designed to direct the condensation back into the flue pipe, where it will be rewarmed and will evaporate, being transported back up the flue again. Within a short time after lighting, if the appliance is run at a high combustion rate, the flue will be warm enough to prevent the condensation occurring and so the moisture will leave the top of the chimney as water vapour. The illustration here shows how a sump connector should be located and how the condensates are directed back into the flue pipe.

Back Outlet Stoves

Much more consideration has to be given to this method of installation to ensure adequate cleaning access, a large enough hearth and sufficient clearance from walls and so on. The first important point is that regardless of what fuel is burned in the stove, the

flue pipe out of the back of the stove is only allowed to travel horizontally for 150mm (6in) before it has to change direction to travel upwards. In most cases, this will be a 90-degree bend to take the flue vertically, but if the flue has to go back a greater distance than 150mm before it lines up with the chimney, then it can go up at 45 degrees for a distance before turning to the vertical.

In the case of a solid fuel stove it is not legal to use a 90-degree bend on the back of the stove, or at the point where the flue turns to the vertical, as there must be a debris collection space created. This requires a tee piece to be fitted at this point. The illustration here shows a back outlet appliance installed in a recess, with a tee piece making it suitable for a solid fuel appliance.

As can be seen from the illustrations of this type of installation the appliance will stand proud of the recess and so will have a greater movement of air around it, but this will also mean a larger hearth area and a greater intrusion into the room.

138

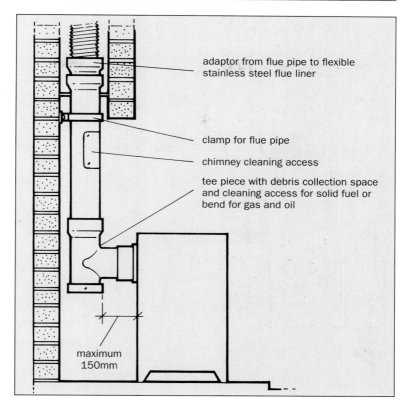

Fig 116 A back outlet stove in a recess showing the required debris collection space and maximum horizontal length.

adaptor from flue pipe to flexible stainless steel flue liner

clamp for flue pipe

chimney cleaning access

tee piece with debris collection space and cleaning access for solid fuel or bend for gas and oil

maximum 150mm

Free-Standing Without a Recess

This type of appliance consists of a stove or room heater placed either in front of a chimney breast or directly under a prefabricated chimney system, and so there is no recess or enclosure for it to stand in.

Some appliances, including some solid fuel ones, do not need to have a constructional hearth, which will be explained in more detail below. For those appliances that do require a constructional hearth the minimum size is 840 × 840mm (32¾in) and the thickness must again be 125mm (5in). It is important to recognize that this minimum size will limit the size of stove that can be installed on it as there must be at least 150mm (6in) of constructional hearth either side of the stove, and also at the back if it is not up against a wall, and at least 225mm (9¾in) in front of the appliance, or 300mm (11¾in) if it can be used as an open fire.

Fig 117 A free-standing stove against a wall, with a prefabricated chimney system.

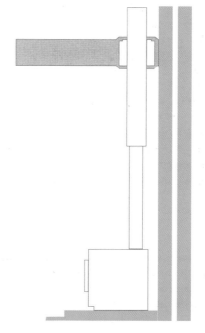

with factory-made insulated chimney

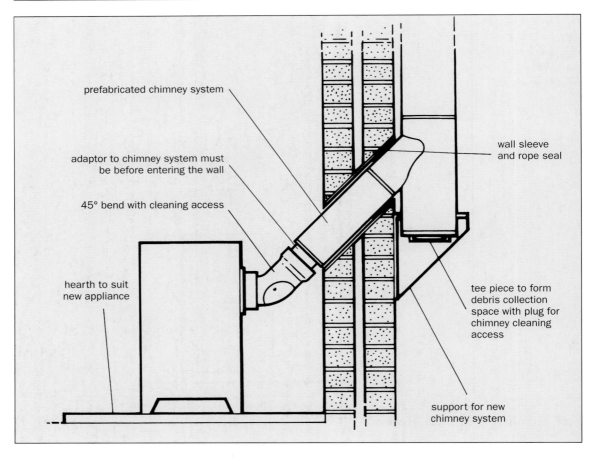

Fig 118   A back outlet stove against an outside wall using a prefabricated chimney system.

It is a fact that where there is no recess present most of the gas and oil-fired stoves will not require a constructional hearth, nor will a number of solid fuel appliances. If the appliance is not capable of producing a temperature of 100°C or more on the hearth then it can be installed without a constructional hearth. In this case, the appliance only needs to be stood on a non-combustible board or tile of 12mm (½in) thick. This board can be placed directly onto a wooden floor or other combustible floor surface. Because this type of installation does not require a constructional hearth, which means that the floor does not have to be taken up for one to be fitted, it has become increasingly popular.

This type of installation can again be split up into sections, and the first one we will deal with is where the appliance is stood against a wall and has a prefabricated chimney or flue system to which to connect.

Figure 117 shows a diagrammatic representation of this type and this is often the easiest way to install a new stove in a house without a chimney as it will only require minimal intrusion and no structural or foundation work.

Sometimes, the size of the property makes it difficult to take the chimney system through the bedroom above because it would take up too much space in that room. However, these chimney systems have been designed to enable the flue to be taken

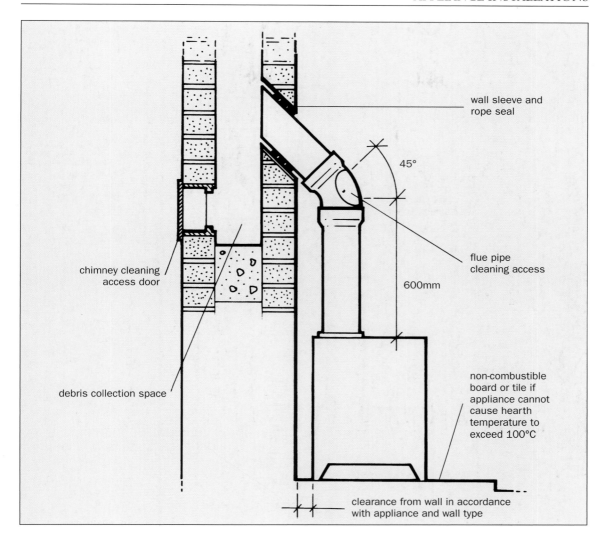

Fig 119 A multi-fuel stove in front of a chimney breast, using a top outlet.

through the outside wall and up the outside of the property, although it must be remembered that planning consent may be required.

It may be that the appliance can stand in front of an existing chimney breast and have the flue pipe taken into this from the appliance. If a stove has the flue coming off the top of the appliance and then taken back into the chimney breast behind it, there are some important considerations. It is important to ensure that you conform to the installation instruc-

tions for the appliance, but below are some of the factors to consider when contemplating installing a multi-fuel or wood-burning stove of this type.

- The flue pipe must go vertically for 600mm (23in) before any bend.
- Any bend in the flue pipe must be no more than 45 degrees from the vertical.
- The bend in the flue pipe must have a cleaning eye on it.

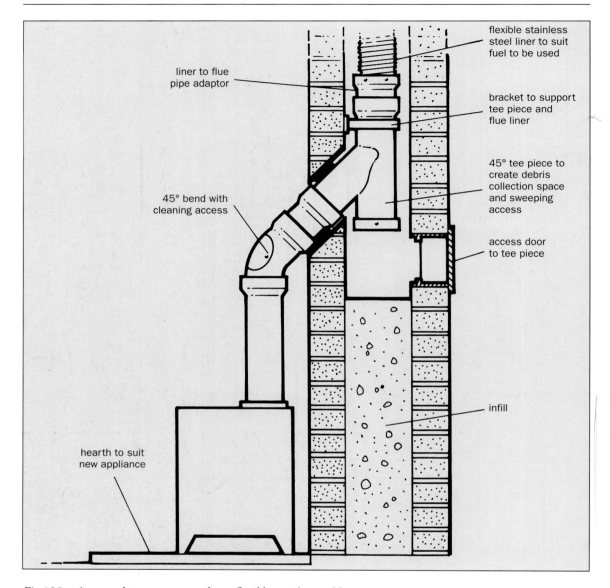

Fig 120   A top outlet stove connected to a flexible stainless steel liner.

- The flue pipe must be sleeved where it passes through the chimney breast wall.
- The flue pipe must be sealed with a fireproof rope seal where it passes through the wall sleeve.
- The flue pipe should not extend beyond the inside of the chimney breast wall and must be cut to match the angle of the inside wall of the chimney breast.

- There must be a debris collection space below the point where the flue pipe passes through the wall.
- There must be an access door, which is sealed and insulated, to enable the cleaning of the debris collection space and the sweeping of the flue.

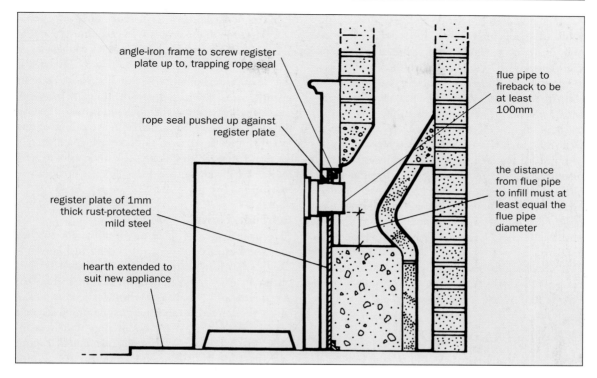

Fig 121   A back outlet stove standing in front of an existing open fire with the fireback still in place.

Labels for Fig 121:
- angle-iron frame to screw register plate up to, trapping rope seal
- rope seal pushed up against register plate
- register plate of 1mm thick rust-protected mild steel
- hearth extended to suit new appliance
- flue pipe to fireback to be at least 100mm
- the distance from flue pipe to infill must at least equal the flue pipe diameter

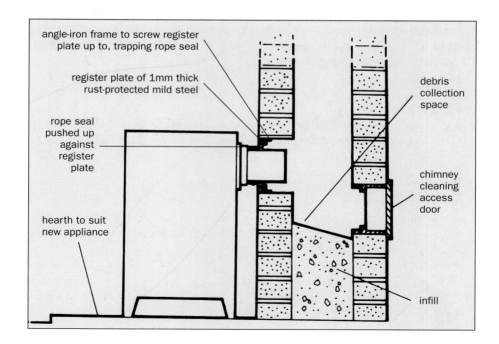

Fig 122   A back outlet stove standing in front of a chimney breast with the open fire removed.

Labels for Fig 122:
- angle-iron frame to screw register plate up to, trapping rope seal
- register plate of 1mm thick rust-protected mild steel
- rope seal pushed up against register plate
- hearth to suit new appliance
- debris collection space
- chimney cleaning access door
- infill

143

Another major consideration is the position of the appliance in relation to any wall behind or at the side of it. If the wall is 50mm (2in) or less from the appliance it must be 200mm (8in)-thick solid non-combustible material for a height of 1.2m (47in) above the hearth, or 300mm (11¾in) above the appliance, whichever is the greater. If the wall is 51–300mm (2–11¾in) away from the appliance the wall must be at least 75mm (3in)-thick solid non-combustible material for the 1.2m above the hearth or 300mm above the appliance. If the appliance is more than 300mm away from the wall then its method of construction is not an issue, but we must not forget the clearance required around the flue pipe, as this could be greater than the 300mm for the appliance.

If a back outlet stove is used in front of an existing chimney breast the installation should be as shown in Figures 119–122.

COOKERS AND BOILERS

It is obvious that normal gas and electric cookers do not need a chimney and so these do not need to be dealt with here, but a solid fuel cooker or boiler will. In nearly all cases these appliances should be considered in the same way as a solid fuel stove, again depending on whether they are fitted in a fireplace recess or stood in front of a chimney breast. Some of the special considerations required in a kitchen are discussed below.

A Hearth

It is important to ensure that the appliance is on a constructional hearth if it requires one, but special consideration to the decorative hearth must be given. To put a decorative hearth in front of a cooker can be very dangerous because people could trip over it when carrying boiling water or hot pans. For this reason, it is common practice to mark the safe perimeter in some way that does not involve a change in level. This may be a change in the colour of any ceramic tiles used as a floor covering, or even a change in the grout colour. It is important to gain building control approval for whichever method you choose.

Kitchen Units

It is common, but incorrect, practice for kitchen units to be placed close up to a solid fuel cooker – this is in fact dangerous. Some cooker manufacturers give a recommended clearance that can be as little as 100mm (4in), but whatever it is this must be regarded as the absolute minimum clearance from the units. Even then, it is important to consider the clearance of combustible material from the flue pipe if a single skin, uninsulated flue pipe is used.

Extract Fans

In a kitchen it is common to have an extract fan fitted. This may be in the form of a cooker hood or similar, but this should not be done with a solid fuel appliance as the fan may be powerful enough to draw fumes and smoke from the flue. If you are requested by the building control officer to fit an extract fan in the same room as the solid fuel appliance, contact HETAS who will be able to give advice.

CHAPTER 10

Testing and Commissioning

CHIMNEY SWEEPING CHECK

This check is carried out for two reasons: one is to check that the whole of the flue can be swept and to help to decide if any additional sweeping-access places need to be added as a part of the work being carried out; and the other is to clean the chimney well before the next two tests are carried out. As with all things related to flues and chimneys, even the simple act of sweeping a chimney can be critical and should be left to the professionals such as HETAS-registered sweeps or members of the National Association of Chimney Sweeps (NACS) or the Guild of Master Sweeps. To be sure of cleaning a chimney, not just pushing a brush through, the sweep will have to make decisions such as which grade of rods to use and which size, stiffness and material of brush to use to ensure that the maximum amount of deposits are removed.

What is being checked for in this test is that the flue is not blocked and that it does not take a route that will allow a large amount of deposits to build up in any given area. By the 'feel of the rods' as they pass up the flue a professional chimney sweep will be able to determine if there are any restrictions in the flue or any sharp changes in direction, all of which will help in the assessment. If a professional sweep is used, he should issue a certificate stating how easy or difficult the process was and if he dislodged any material that indicated faults in the flue and so on, as well as advice on the best place for any additional cleaning access holes that may be required.

CORE BALL TEST

With the chimney sweeping test complete it should now be safe to carry out a core ball test. A 'core ball' is a heavy ball, often concrete or cast iron, which is lowered down the flue from top to bottom on the end of a rope, hence the need to sweep the chimney first. The ball should be between 12–25mm (½–1in) smaller than the size of flue required for the appliance and should be as round as possible (ensure that the opening at the base of the flue is adequately closed off and sealed as more soot may still be dislodged).

When a sweep's brush passes up a flue it can be pushed through a hole as small as 100mm (4in) in diameter and so the fact that the sweep has got the brush from top to bottom is no guarantee that the flue does not restrict down to below the size of flue required for the appliance to be installed. On the other hand, if it is possible to pass a solid concrete ball of the required size down the flue, it proves that the flue is of the minimum area at all points.

In addition to this, if the ball can pass through the flue under gravity alone, then the angle of any bends in the flue cannot be too steep or the ball would just sit on the horizontal section (the ball cannot roll down the slope, it has to slide because the rope prevents the rolling action). Once the ball has passed from top to bottom under its own weight then it is safe to pass on to the next test.

AIR-TIGHTNESS TEST

The importance of a flue being reasonably airtight cannot be overstated. Leaks in a flue will allow cool air to enter and reduce the flue gas temperature (and

145

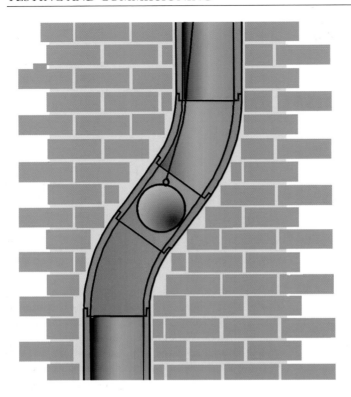

hence the draw), increase the risk of deposits and acid condensation, as well as eventually reaching a point where smoke and fumes may leak out into the house.

The basis of this test is laid down in the British Standards and repeated in the *ADJ* and is given below:

- The appliance must not be alight at the time of the test.
- Before commencing, the chimney flue must be warmed for at least ten minutes using a blowtorch or a single burner camping gas stove (the latter is better as it directs the heat up into the flue and normally has a stable stand or base on which it can be seated). For large or tall flues this may require longer.
- The opening to the fire must be closed up with polythene or paper and masking tape to prevent the loss of smoke.
- Smoke must be introduced into the flue using smoke pellets and the amount will depend upon the overall volume of the flue.

When smoke is observed coming from the chimney pot this has to be sealed. This can be done using either the closed cell foam bungs that are available in various sizes, or by using a polythene bag (ensuring there are no holes in it) placed over the pot and sealed in place with masking tape.

The chimney stack, chimney breast and surrounding walls must then be closely studied at ground level, first floor level, roof space level and terminal level to check for smoke leakage. Particular care must be taken when observing walls outside as any breeze may take the smoke away quickly and it can be difficult to see in bright sunlight or against grey skies. As leakage may occur some distance from the original fault particular attention must be taken at barge overhangs, window reveals and around any pipes that may pass through a cavity wall as well as from any other flues that are in the same chimney stack.

The test should last for at least five minutes and the skill of the tester is in identifying whether or not there has been any significant leakage when the bung or bag is removed from the top of the chimney pot. If

Flue Size to Smoke Pellet Ratio Based on a Smoke Pellet Producing at Least 18m³ of Smoke	
Volume of Flue (m³)	Number of Smoke Pellets
Up to 0.1	1
0.1–0.2	2
0.2–0.4	3
0.4–0.6	4

The smoke pellets will have to be of a type that provides at least 18m³ of smoke to ensure adequate volume of smoke. If the amount of smoke used in the flue test is inadequate then the density of smoke may be insufficient to make a leak visible. When smoke starts to form, the fireplace closure is sealed and the smoke allowed to drift upwards. In the case of a closed appliance, such as a multi-fuel stove or solid fuel boiler, the firedoor, ashpit door and any air-control inlets must be closed.

there appears to be less smoke than expected, then leakage may have occurred but not been visible. If this happens, the test should be repeated until the location of the leakage is found.

Approved Document J states that all flues can be expected to leak to an extent, and so some smoke leakage may be seen during the smoke test. It therefore becomes a matter of expert judgment as to whether the level of leakage indicates failure. In these circumstances, it can be very difficult to know what is allowable leakage and the information given in *ADJ* is not much help, in that it states 'wisps of smoke on the outside of the chimney or near joints between chimney sections do not necessarily indicate a fault. If, however, forceful plumes or large volumes of smoke are seen, this could indicate a major fault.' In my opinion it is the case that if you can see smoke leaking from a flue, especially a masonry chimney, no appliance should be lit until remedial action has been taken and a successful test completed.

PRESSURE TEST TO EUROPEAN STANDARDS

This test is an alternative to the air-tightness test above and although it is more definitive in deciding on the level of leakage from a flue, and can state whether it is within the permissible level or not, it will not show where the leakage is or whether it is from one large leak or from hundreds of small ones. If the test indicates the flue has failed or is close to the failure level, the smoke test above will be needed to

identify what type of leak exists and what remedial action is required.

The basis of this test is that the top of the flue is bunged with a closed cell bung and the bottom has an air-delivery hose sealed in place. The instrument used for the test will then deliver air into the flue until the test pressure is reached, in the case of the N2 test this is 20pa (*see* Chapter 5). The machine then delivers just enough air to maintain this pressure and will take three measurements of the air movement needed to achieve this. The air being delivered will equal the air leaking out of the flue if the pressure is maintained, and so the measurements taken are equal to the leakage rate from the flue. The operator then enters the diameter, or rectangular size, of the flue and the effective flue height into the instrument, which will then calculate how much air is leaking out per m² of flue surface area; the allowable rate for an N2 test is 7.2 m³/hour/m² of flue surface area. The instrument will then give a pass or fail report that can be printed out as evidence of the test and the result.

This is a more definitive test, which is both reproducible and not so dependent on the opinions of the tester.

EVACUATION OR SPILLAGE TEST

This is usually the last test to be carried out. If it is part of the commissioning of the appliance or is intended to leave an existing appliance in use, then it should also be the final test before leaving the property.

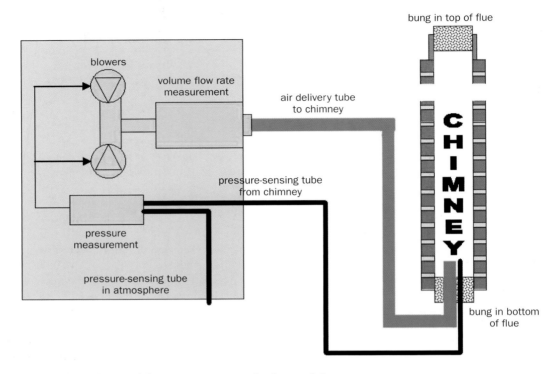

Fig 124   Diagramatic layout of the pressure test set-up for flues and chimneys.

The test will vary according to the type of appliance to be installed and the type of fuel to be burned. If a gas appliance is to be used then the appropriate test can only be carried out when the appliance installation is complete; if a solid fuel or multi-fuel stove is to be installed the same applies. If the opening is to be used as it exists for an open fire, of any fuel type, then this test is designed to ensure that any smoke or fumes produced by the fire are drawn into the flue and taken out to the atmosphere.

Before starting this test, the flue must be warmed for a period of at least ten minutes as described for the air-tightness test above.

All of the doors and windows in the room should be closed, as should any closable vents such as those in window headers and so on. If there is an extract fan in the same room as the appliance, or a ceiling fan, this should be switched on during the test.

A smoke pellet of the type used for the air-tightness test should be lit and placed on the forward-most part of the grate.

The smoke pellet should be allowed to burn halfway through and at all times the smoke should enter the chimney and none should escape into the room. At the halfway stage the operative should go outside and check that the smoke is only leaving the flue through the correct chimney pot and not from any other pots on the chimney or from any joints in the chimney stack or gable end wall. The operative should then return to the room containing the appliance and ensure there is no smell of test smoke in the room. If a smell is present this indicates spillage and so the appliance should not be used until the reason for this has been identified and corrected (*see* Chapter 11).

In the case of a stove or other boiler, this test should be carried out in the same way, except that the appliance door should be closed and all air inlets opened. However, if the manufacturer states that the appliance can be used with the appliance doors open, the test must be carried out with these allowable openings fully open.

CHAPTER 11

Care and Maintenance

As with all mechanical equipment and building components, a heating appliance, its fireplace and its chimney will require regular servicing and maintenance to ensure their safety and stability. Many of these appliances will not require any servicing or maintenance to be carried out by the user; for example, if they are gas or oil-fired all maintenance work must be done by either a CORGI or OFTEC-registered engineer.

In the case of an open-flued appliance, such as a DFE gas fire, consideration should be given to having the chimney swept every year before it is used again in the autumn. This is not because there may be a large build-up of soot in the chimney, but because we do not know what may have happened in the chimney since the fire was last used in the winter or spring. In older properties the flue may have become damaged or masonry may have fallen and partially blocked the flue or caused a serious leak. In any property a bird may have nested in the flue (or even a squirrel in some parts of the UK). Larger birds such as a jackdaw, raven, or even a gull, can block a flue in days! If a jackdaw decides to nest in a flue it can create a blockage of over a metre in depth in less than a week, and so it is wise to have the flue swept after the nesting season but certainly before using the fire in the autumn.

In the case of a solid fuel appliance the chimney must be swept at least once a year irrespective of how often it has been used. If deposits are allowed to build up year after year they can become very hard, and removal at a later date could be quite costly. If a solid fuel fire is used often and burns wood or coal, then it would be wise to have it swept halfway through the heating season as well.

The sweeping of a chimney is not intended just to prove that a flue is not blocked, but also to clean the sides of the flue to avoid a gradual build up of deposits over the years. To clean a flue while it is being swept, the sweep must ensure that he chooses the correct size and stiffness of brush to suit the size of flue, the material it is constructed from or lined with, and the fuel that has been burned since it was last cleaned. A professional chimney sweep should have a great number of brushes in his equipment as well as different strengths of rods. Once the chimney has been swept the chimney sweep should issue a certificate stating whether, in his opinion, the flue is safe for you to continue using and when he estimates the next time the chimney should be swept.

There are several trade associations for chimney sweeps, including the National Association of Chimney Sweeps (NACS), the Guild of Master Sweeps and a register of sweeps maintained by HETAS.

In-between the sweeping of the flue, many solid fuel stoves require the user to carry out routine maintenance in addition to fuelling and the removal of ashes. One of the most important of these is the clearing and cleaning of any flue ways in the stove and any baffle plates that may be present. To improve the efficiency of stoves there is often a baffle plate above the fire to prevent the flames and hot gases going straight into the chimney and to deflect them against areas of the stove that will take heat from the gases and deliver it to the room or to an oven. These plates will need to be removed and cleaned at regular intervals to ensure that the efficiency of the stove is

Fig 125   Sweep's brushes of many types and sizes.

maintained and, more importantly, to ensure that all the gases can enter the chimney without leaking from the appliance.

As with all issues, study the manufacturer's recommendations and follow them carefully; however, this type of flue and baffle cleaning operation is best done every month if the appliance is used on a regular basis, such as through the winter.

Other issues that may apply to your new appliance include the following.

• Regularly check that air vents are clear and not blocked with leaves or even spiders' webs!

• Have the appliance serviced every year.
• If it is a solid fuel stove, use only those fuels recommended by the appliance manufacturer or HETAS. Always buy your fuel from a reputable source, such as a member of the Approved Coal Merchants Scheme (ACMS).
• Have a carbon monoxide alarm fitted in the same room as the appliance.
• Always follow the user instructions when operating the appliance and if operating tools are provided ensure these are in good repair and readily available.

CHAPTER 12

Typical Chimney Problems and their Solutions

FLUE FUNCTION

The updraught or 'pull' in a flue results from a combination of the height of the flue and the difference in temperature between the flue gases and the outside air. Very simply, the column of hot gases in a flue is lighter in weight than an equivalent column of cold air outside, so the pressure inside a warm flue is less than the air pressure outside. It is this quite small difference in pressure that creates updraught. The warmer and taller a flue, the better the draught and the less risk of condensation.

Large voids or pockets, rough surfaces and sharp bends or steep angles in a flue are resistances to the flow of flue gases and so reduce the draught. For this reason, they should be eliminated as far as possible, particularly in the case of open fires.

Air leakage into a flue has the effect of cooling the flue gases and therefore reducing draught. In certain circumstances, it may be desirable to reduce an excessive draught by introducing 'dilution' air into a flue, but this should only be done in a controlled way, not by allowing accidental air leakage.

Wind at the top of a chimney may have a positive or negative effect on draught, depending on the surroundings and the position of the chimney top in relation to the roof.

Diagnosing Problems

Faulty chimney draught may show itself either in a sluggish fire and general poor performance of the appliance, or it may cause the actual escape of smoke and fumes into the room.

A fire will burn sluggishly when the draught is too little to supply sufficient air to the fire but there is just enough chimney 'pull' to carry away the smoke; this usually occurs with independent boilers, cookers and room heaters. On the other hand, the opposite problem of too much updraught sometimes arises with closed appliances, such as independent boilers. This makes it difficult to control the burning rate with the normal combustion air control, particularly in windy weather.

When smoke or fumes actually escape into the room, the more usual reasons are:

- Symptom A: there is insufficient draught to carry away all the smoke.
- Symptom B: there is no updraught.
- Symptom C: the chimney terminates in a high-pressure zone.
- Symptom D: there is a downdraught and difficult site conditions.

The smokiness can be due to a variety of causes, several of which could operate together. It is therefore not always a simple matter to diagnose the cause of the problem, nor is there any universally successful treatment or 'cure-all' – each individual case has to be treated according to its own symptoms.

To be able to cure smokiness, it is first of all necessary to know why it occurs, and because of the various possible reasons, not all of which are immediately obvious, it is advisable to set about the job in an orderly way and to follow a set sequence of operations.

The first step is to read through this section, as sometimes a fault can be recognized immediately from a description of it and the trouble corrected without the need for a lengthy investigation. If, however, the cause of the trouble is not obvious, the next step is to check the symptoms, the more common of which are detailed below.

Symptom A

Some smoke or fumes, but not all, escape into the room without any sign of being blown back by the wind. Smokiness is constant and usually occurs regardless of weather conditions, although the wind may have some effect in either worsening or improving matters. This usually means insufficient draught.

Symptom B

There is no updraught at all, but no sign of blowing back. The problem is usually constant and unaffected by wind conditions.

Symptom C

With the wind in a certain quarter the draught stops or reverses and causes smoke emission; the draught returns when the wind ceases or changes direction.

Symptom D

There is intermittent blowing back with the wind in a certain quarter. The degree of smokiness will vary with wind strength.

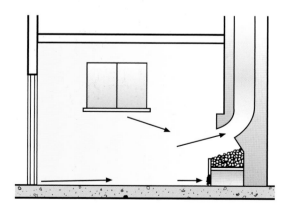

Fig 126 Air starvation is when the air into the room is insufficient to drag all the smoke into the flue.

Finally, one or more easy tests should be carried out, the results of which will usually indicate the cause of smokiness. It is not unknown for two or perhaps three different tests to give positive results; this may mean that there are several causes operating together. For example, a simple case of downdraught could be complicated by air starvation, or a badly formed throat could be associated with a partially blocked flue.

For quick reference, the possible causes of smokiness, methods of testing and suggested remedies are listed as A.1, A.2, B.1, C.1, and so on, to cross-reference with the main symptoms outlined above.

Too much chimney draught and condensation in a flue are dealt with under separate headings at the end of this chapter.

Symptom A: Insufficient Draught

A.1. Air Starvation

All solid-fuel-burning appliances need a flow of air into the room; some, particularly the open fire, need more air than others. In addition to the air required to burn the fuel, a much larger quantity flows over the fire, through the appliance or fireplace opening, and up the flue.

A closed appliance may need only 15–25m³ (530–883ft³) of air per hour, whereas an inset open fire with a large fire opening and a large 'throat' induces the flow of an extra 260m³ (9,180ft³) or more per hour. If the required amount is not available, the air speed through the fireplace opening is reduced and may fail to carry all the smoke into the flue. Sufficient air to carry away the smoke cannot enter a room that is too well draught-proofed, or which has a solid concrete floor and well-fitting doors and windows.

Test Open a door or window, preferably when there is no wind to complicate matters. If smoking into the room ceases, the trouble is usually due to air starvation.

Remedy An air vent of the correct size should always be present as described in Chapter 2, but it may be necessary to allow more air into the room. However, achieving this in a way that also avoids unpleasant draughts is often difficult. The first step

BAD EXAMPLE

BAD EXAMPLE

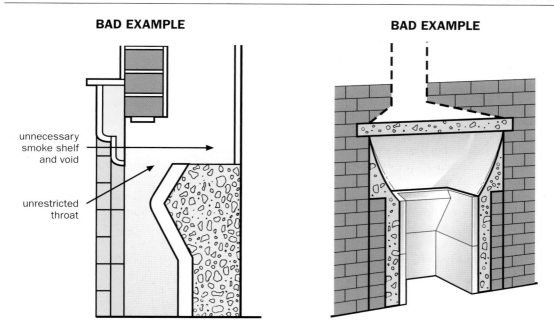

unnecessary smoke shelf and void

unrestricted throat

Fig 127   A large throat or void can increase the demand for air, causing air starvation even if a vent is provided.

GOOD EXAMPLE

GOOD EXAMPLE

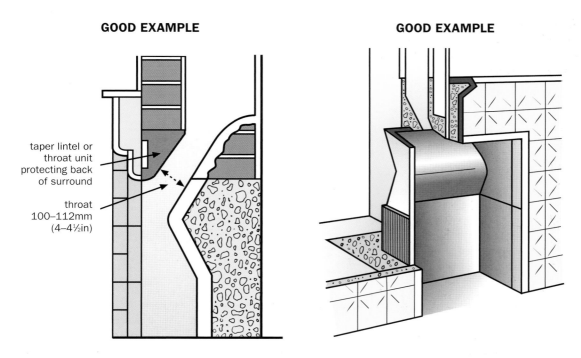

taper lintel or throat unit protecting back of surround

throat 100–112mm (4–4½in)

Fig 128   A streamlined throat and gather aid air flow but reduce air demand.

should be to see if it is possible to reduce the volume of air flowing up the flue so that the 'demand' for air is less. An inspection of the throat over the fire may reveal a large throat without any attempt at streamlining and an even larger void beyond and at the sides.

The void should be filled in and the throat reduced so that there is a smooth, streamlined entry into the flue. If the throat opening can be made adjustable for size it will be better still, as then it will be possible to find the minimum amount of opening to suit the 'pull' of the flue when the chimney is warm, and yet allow the full opening for relighting the fire when the chimney is cold.

There are proprietary throat restrictors that can be easily fitted, but these must never be used with a gas appliance. When these units are used it is possible to reduce airflow from the room into the flue. This reduction, in conjunction with perhaps the removal of some over-zealous draught-proofing around doors and windows, may cure the less serious cases. Bad cases often require further treatment. As well as helping air starvation, a throat restrictor is an excellent means of making a room more comfortable by reducing the air induced up the flue and so reducing excessive ventilation (*see* Figure 131).

Additional ventilation for the room to satisfy the 'demand' caused by the pull of the flue can usually be provided without resulting in unpleasant cold draughts. Fitting a draught-master type of vent over the door to the room will draw air from the hall, which is less cold than the air outside.

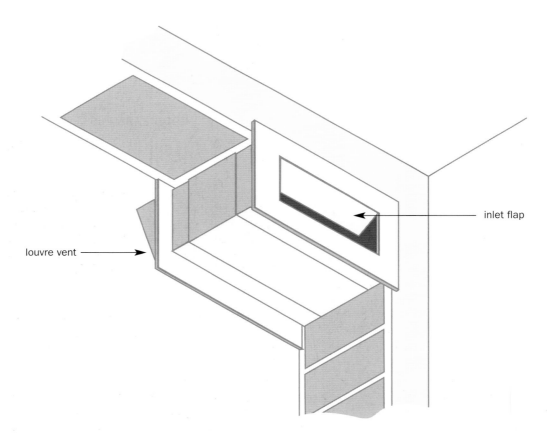

inlet flap

louvre vent

Fig 129   Draught-master vent fitted over a door between the hall and the room containing the fire. This has a mica flap to provide fire protection and to allow only the air required by the fire into the room.

Fig 130   A floor vent with a brass grille may not provide adequate air on its own but can make a useful contribution. There are not to be used with gas appliances.

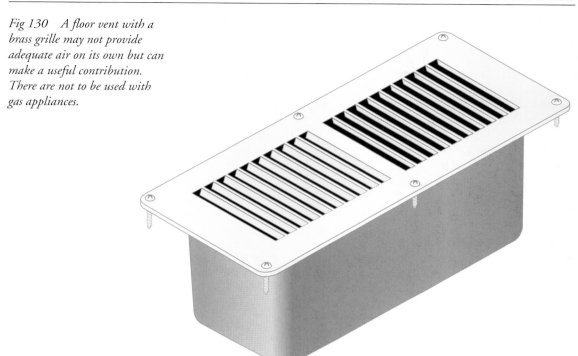

With a suspended floor (boards on joists), extra air inlets fixed in the floorboards adjacent to the hearth may sometimes help, but the results are frequently disappointing unless they are large enough. Where the floors are solid, extra air inlets are necessarily limited to the walls.

The most effective way of dealing with the problem of air starvation is to install a modern closed appliance in place of the open fire; this drastically reduces the volume of air from the room flowing up the flue and increases heating efficiency.

A.2 Throat over Fire Too Large or Badly Formed

The velocity of the gases at this point should be high, so that the action is similar to that of a vacuum cleaner nozzle. The smooth, swift movement of the smoke and gases may be impeded by the irregular shape of the throat and the large void above, at the sides or behind it. In the previous section the illustrations show a bad throat and void construction, plus the proper reconstruction of this throat and stream-lining of the gather up to the point where the flue itself begins.

Test   Very often the turbulence and eddying of smoke can be clearly seen. A simple test makes use of a piece of cardboard or thin sheet metal cut and bent so that it can easily be wedged across tile front of the throat (in a similar position and shape to the stream-lined lintel shown in the illustration in the previous section).

Remedy   Reconstruct the throat as shown, or, perhaps better still, fill in and streamline the voids and fix an adjustable throat restrictor of the type shown in the illustration below. Filling in, which should be carried up as near to the flue as possible, is usually done with bricks and rubble bedded in lime mortar and overlaid with a lime mortar rendering. But remember that mortar will not bond to a sooty surface and the existing rendering should be well scraped first.

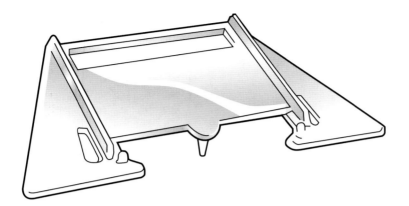

Fig 131 An adjustable throat restrictor for solid fuel open fires.

A.3 Fireplace Opening Too Large

The flow of air and other gases through a flue is limited by the size and constructional details of that flue. All flues have a particular volume-carrying capacity or rate of flow according to conditions. If this same amount of gases has to flow through a larger area, such as the fireplace opening, the velocity will be reduced and the result may be that smoke will eddy into the room before it can be drawn up the flue.

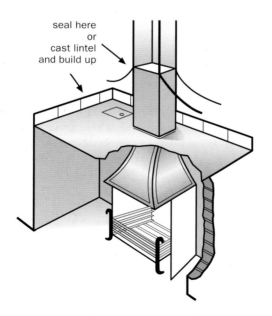

seal here
or
cast lintel
and build up

Fig 132 Using a canopy to change an effective fireplace opening that is too large for the flue.

Usually an inset open fire with a conventional opening of 550 × 400mm (21½ × 16in) requires a flue of not less than 200mm (8in) diameter, or 185mm (7½in) square, to clear the smoke and gases. Where the fireplace opening is too large in relation to the flue there is a likelihood of smokiness, particularly if there is a badly formed throat. This trouble is very common with dog grates or basket grates standing in very large openings.

Test There is no simple test for this condition; all that can be done is to check the size of the flue and compare it with the size of the opening. If the area of the fireplace opening is more than eight times the area of the flue, there is a likelihood of smokiness. With a short flue, as in a bungalow or top floor flat, this ratio may need to be reduced to 6:1. There are, however, many variable factors and it is difficult to lay down hard and fast rules, but the graph from the Solid Fuel Association will help.

Remedy Altering the size of the flue is usually an expensive option; it is also often impractical to reduce the fireplace opening. All that can be done with an inset fire is to consider replacing it with a closed room heater. If you wish to retain a dog-grate or similar fire in a recess in a period setting, smoking back can often be overcome. This is done by fixing a metal register plate across the underside of the top of the recess, and a metal canopy over the fire, with its constricted end projecting through the register plate. This should go as high as possible to connect to the flue gather.

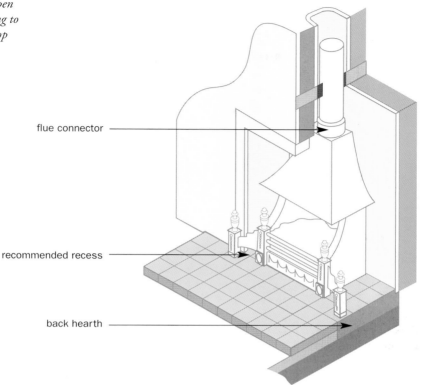

Fig 133   A free-standing open fire fitted into a large opening to match the flue size and to stop smoking-back.

flue connector

recommended recess

back hearth

Better still, however, would be to install a purpose-made free-standing fire, which also has an integral throat restrictor. This is enclosed at the back and two sides up to the canopy. All the over-fire air for the flue must pass through the fire opening, which is properly related in size to a 225 × 225mm (9 × 9in), or an equivalent 250mm (9¾in), diameter flue. Such a fire can be positioned in a large recess without smoke-back occurring, providing that the height of the recess is not greater than the outlet end of the canopy and the flue is properly constructed.

A.4 Fireplace (Surround) Opening Too High
Although modern surrounds often have 550mm (21½in) high openings (for raised fires), or 500mm (19½in) (for hearth-level fires), there are still many open fires with 600mm (23in) openings. The higher the opening, the easier it is for smoke to trickle out into the room before reaching the flue, particularly if there are aggravating circumstances such as cross-

currents of air, due perhaps to doors on both sides of the fireplace, or if there is a restricted flow of air into the room.

Test   Cut a piece of cardboard or similar material a little longer than the width of the opening and about 100–150mm (4–6in) wide. Place this across the top of the opening so that the latter is reduced in height to 550mm (21½in) from the hearth, or 500mm (19½in) for a sunken fire.

Remedy   If smoking ceases, a permanent non-combustible canopy or plate securely fixed in the same position is the simplest cure, but ensure that the throat is also properly constructed.

A.5 Flue Offset Too Low, Too Abrupt or Too Long
Flues are very frequently carried over, or 'offset', to one side of the chimney breast, either to make room for other flues in the same stack, or to bypass the

157

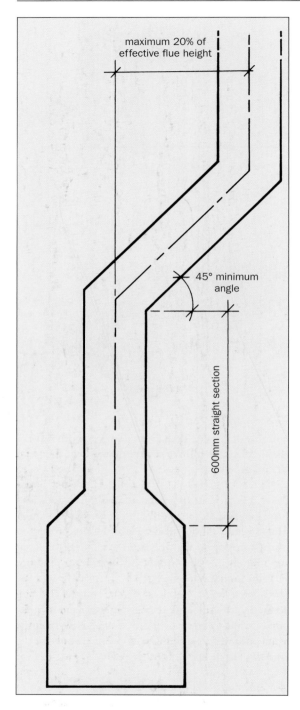

maximum 20% of
effective flue height

45° minimum
angle

600mm straight section

Fig 134   The features of an offset in a flue that has been well designed.

fireplace upstairs. Sometimes this offset is too low, or the bend is too abrupt, or perhaps the traversing length is too long – sometimes all three faults are found in the one offset. The result is poor chimney draught and a smoky fire. An offset should ideally start no less than 600mm (23in) from the top of the fireplace or stove, it should be at least 45 degrees from the horizontal and be no longer than 20 per cent of the effective flue height.

Test   It is not always an easy matter to follow the run of a flue, and probably the easiest means of checking is with a chimney sweep's rods. These will indicate the height of the first bend and give some idea of the angle of traverse. A long traverse is only possible in a wide chimney breast or in an unusual chimney construction.

Remedy   There is usually no alternative to opening the front of the chimney breast (or back, if it is on an outside wall) and rebuilding the offending part of the flue. It is such a major operation, however, that all other possible causes of smokiness should be checked first. The essential features of a good offset are shown in the illustration here.

A.6 Partial Blockage or Constriction of the Flue
This usually occurs at an offset and may be due either to bad workmanship at that point, or foreign matter or soot accumulation at the bend. A common cause of trouble is mortar that was dropped down the flue when it was being built and now partially blocks the bend.

Test   The sweep's brush will often reveal such obstructions, and may sometimes dislodge them if a scraper is fitted to the end of the rods. Another method of checking a flue for obstructions is to lower a 'core ball' on the end of a rope down the flue. The core ball, made of concrete or metal, should be of such a size that it has not more than 25mm (1in) clearance in the flue. In order to determine the exact position of the blockage, a knot is tied in the rope where the ball stops and the position of the knot in the rope measured to see how far down the flue the blockage is. It may even be possible to dislodge the blockage using the ball.

Fig 135   A partial blockage in a flue caused by soot or mortar deposit will restrict gas flow and cause some smoke-back into the room.

9in × 9in flue with parging or flue liners
showing partial blockage at bend

Remedy   If the blockage cannot be removed by other means, the flue at this point will have to be opened for clearing, and then made good once the obstruction has been removed.

A.7 Unsuitable Size of Flue
The conventional 225 × 225mm (9¾ × 9¾in) brick flue, although perhaps not ideal, is generally suitable for most types of domestic appliances unless they are particularly efficient. Anything over this size, such as the large flues sometimes found in older houses, may cause smokiness or spillage chiefly because such a large flue never really gets warm. On the other hand, flues must not be too small; open fires in general should not have flues smaller than about 200mm (8in) diameter or 185mm (7½in) square for solid fuel, particularly if house coal is to be burned, and 175mm (6⅞in) for DFE gas appliances. Flues of about 125mm (5in)

diameter may well be better for small independent boilers, room heaters and cookers burning smokeless fuel, gas or oil (150mm for multi-fuel appliances).

Test   Check the size of the flue.

Remedy   A very good method of reducing the size of a flue is to have it lined with a flexible stainless steel liner (except for solid fuel central heating boilers), as this is non-permanent and so can be removed later if the original flue size is required again. Other lining methods can be used but tend to be permanent. However, if the flue is also found to leak this is not a suitable method to use and one of the alternatives will have to be chosen. With an undersize flue it is often more practical to change the appliance for a more efficient type, requiring a smaller flue diameter.

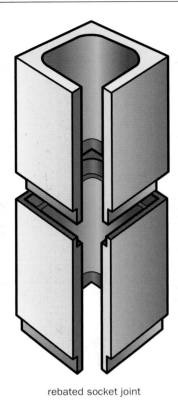

rebated socket joint

Fig 136   Rebated and socketed liners used in modern masonry flues must be fitted correctly with the socket uppermost and well-aligned joints.

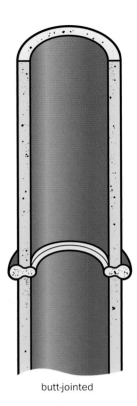

butt-jointed

Fig 137   Poor joints in flues can result in cement rings that will increase flue resistance, no jointing or backfill and leaking joints.

A.8 Bad Fixing of Liners

Since the Building Regulations came into force in 1967 clay flue liners with rebated or socketed joints have been used in masonry chimneys. Some older houses, however, may have liners with plain ends that are butt-jointed. Sometimes these were fitted 'out of line' and overlapped, resulting in a series of obstructive ledges that hinder updraught.

Another fault often found with all forms of section liners is a protruding ring of cement at joints that should have been removed during installation. In other cases, liners have been installed one on top of another with no jointing between them, and infilling between the liners and the brickwork omitted. This is sufficient to cause a leak at each joint and spoil the draught; such a flue is unsound.

Test   A scraper on the end of sweeping rods will usually reveal any obstruction. A smoke test may indicate the extent of any leakage. A competent person should carry out this test.

Remedy   If the trouble is due to cement rings, they can often be removed by careful scraping. For overlapping joints (which can be checked if necessary by cutting open the brickwork of the chimney), there is no treatment apart from removing the existing lining and relining by using one of the methods described previously.

A.9 Unsuitable Chimney Pot

In many cases, round-base chimney pots are fitted to square flues by placing small pieces of tile across the

160

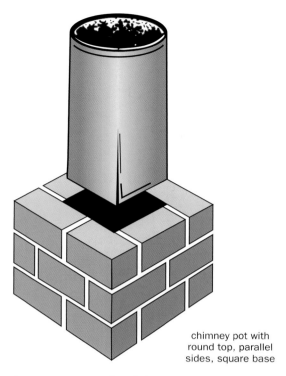

chimney pot with
round top, parallel
sides, square base

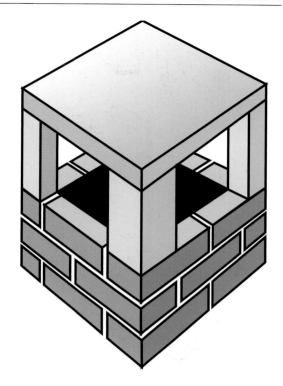

*Fig 138   A square-based chimney pot on a
square flue is always recommended; if the pot
tapers or changes area it must never drop below
the minimum required for the appliance.*

*Fig 139   The detail of a dovecote-type top for
larger flues to protect against rain and bird
ingress.*

corners of the flue to provide a base for the chimney
pot. This results in four obstructive ledges, which,
although not perhaps the sole cause of the problem,
may be the 'last straw' along with other faults. Many
chimney pots taper to the outlet in the belief that this
improves the draught and reduces the entry of rain
and wind. A pot with parallel sides, not tapered, and
with a square base offers less resistance to the outward
flow of products of combustion and is better for an
open fire. This may help to take away the 'last straw'.

Test   By inspection.

Remedy   Replace with a square-based pot with
parallel not tapered sides.

A.10 Partially Blocked Chimney Pot
When smoke flows through a flue, the particles of
soot are particularly attracted to cold surfaces such
as the inside of a chimney pot. Usually most of
the soot is removed by the sweep's brush, but if the
smoke is high in tar content the deposits may
resist everything but a hammer and chisel. In
districts where house coal with a high tar content
or wood is commonly used, chimney pots should be
as short as possible (but not less than 150mm (6in))
and well protected by steep flaunching. It is also
advisable for the pot to have parallel sides with an
outlet not less than 200mm (8in) diameter, or
larger if the flue is larger, and preferably salt-glazed
internally.

161

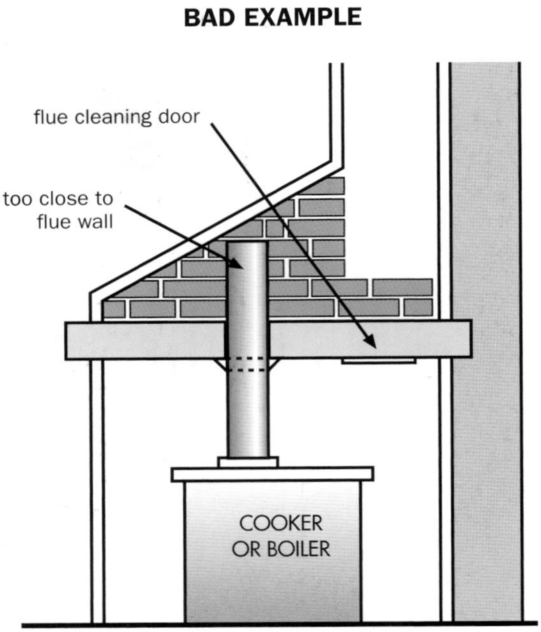

BAD EXAMPLE **GOOD EXAMPLE**

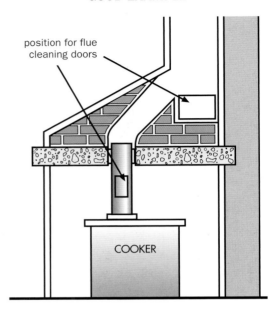

Fig 140   A cooker installed with a register plate and flue pipe into a large void with inadequate clearance and cleaning facilities.

Fig 141   A cooker installed with a concrete lintel, flue formation and debris-collection space.

Test   A sweep's brush usually reveals the obstructions.

Remedy   If the trouble is recurrent, replace the pot with a shorter one of a larger size. The 'dovecote' top shown here is sometimes used to protect large flue outlets against wind and rain; it is not, however, intended to replace the recommended square or circular termination, which should be built into the brickwork so that only 50mm (2in) or so projects above the stack. The combined area of the four openings should equal at least twice the area of the flue.

A.11 Baffling of Flue Gases
The illustrations in this section show fixing faults that often result in a poor draught. In each case, the flue pipe projects too far into the flue so that the free flow of gases is hindered.

Test   By inspection.

Remedy   The type of installation shown in the illustration above is best treated by the replacement of the register plate with a precast or cast-in-situ concrete lintel as shown. This enables the flue pipe to be sealed using fibreglass rope that can be caulked into the gap between the flue pipe and the hole in the lintel. The sides of the lintel should be sealed to the fireplace wall from above using mortar. A new flue should be constructed using liners and infill to prevent any voids. A double-seal soot door should be fitted to the face of the flue in the position shown to enable a collection chamber to be formed.

Where a free-standing room heater is placed in front of an existing open fire a closure plate must be fitted, covering the fire opening and sealed to the fire surround. Infill should be placed behind the register plate up to a height as shown in the illustration here.

162

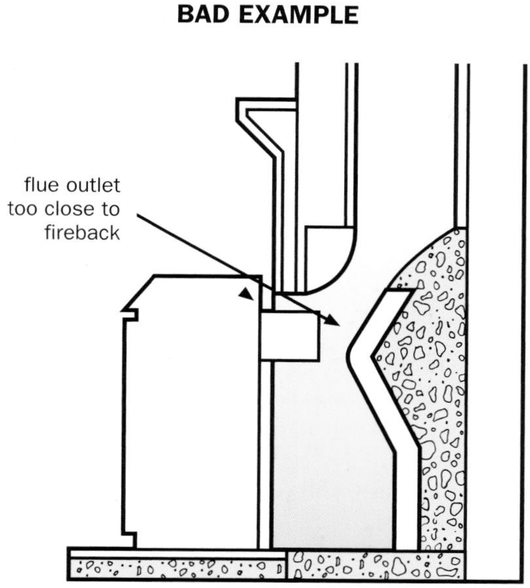

BAD EXAMPLE

flue outlet
too close to
fireback

*Fig 142   A room heater with a back outlet
installed in front of an open fire with long flue
pipe, and an inadequate seal.*

BAD EXAMPLE

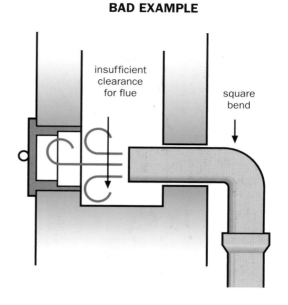

insufficient
clearance
for flue

square
bend

*Fig 144   A flue pipe into a chimney breast with
a square bend, poor seals and insufficient
clearance.*

GOOD EXAMPLE

minimum distance
100mm (4in)

rope seal
between
offtake pipe and
closure plate

minimum
dimension
equal to
diameter
of offtake
pipe

infill
behind
closure
plate

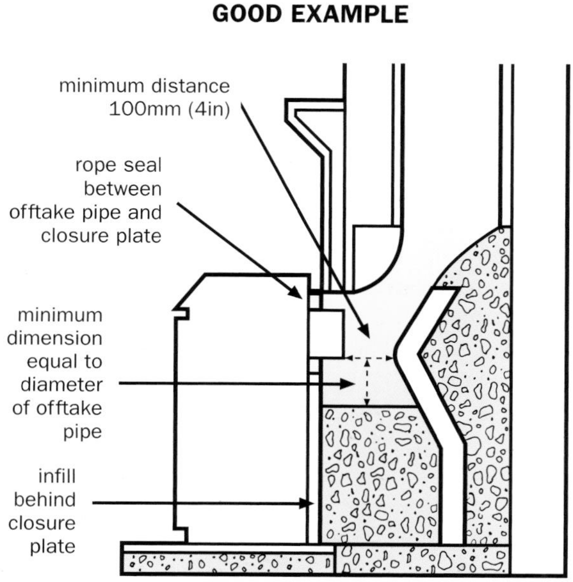

*Fig 143   A room heater with a back outlet installed in
front of an open fire with register plate, seal and infill.*

GOOD EXAMPLE

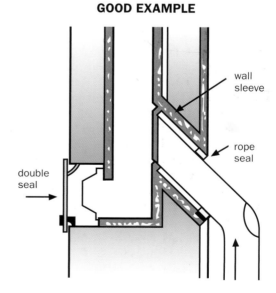

wall
sleeve

rope
seal

double
seal

*Fig 145   A flue pipe into a chimney breast
with a square bend, sleeve, seals and debris-
collection space.*

The flue pipe from the appliance should be sealed into this closure plate where it passes through. This flue pipe must be short enough to allow a clearance of at least 100mm (4in) between the end of the flue pipe and the fireback.

Flue pipes projecting into brick flues should be via bends of 135 degrees to avoid a sharp change in direction of the flue gases, and the pipe should be cut flush with the inner wall of the wall sleeve. To ensure a good seal between the flue pipe and the chimney, a wall sleeve must be fitted and the gap between the flue pipe and sleeve caulked with fireproof rope. The soot door should be 'double' (have an inner plate) as shown in Figure 145 to avoid excessive cooling of the gases. It should also be slightly below the entry of the flue pipe so that the hot gases do not impinge on it and so that there is a small debris-collection space to reduce the possibility of the flue pipe becoming blocked by deposits dropping from the flue. There should be no large cavity below the soot door.

A.12 Air Leaks
Cold air leaking into a flue will naturally cool the gases and reduce draught, and it is most important that flues should be as airtight as possible. The more usual points of air leakage are: around register plates where they butt against the brickwork; the joints around flue pipes projecting through register plates or brickwork; and the joints of flue pipes and soot doors.

It is also possible for a brick chimney to leak air if the brick joints are bad or have cracked, and for air to be drawn from an adjacent disused flue if the brickwork between them (mid-feather or withe) is faulty. In certain conditions, this may also result in potentially dangerous outward leakage of fumes.

Even flues with modern liners will leak unless they are properly jointed during installation with high alumina cement and the space around them solidly filled in with an insulating mix.

Test Inspection will usually reveal leaky joints, and the flame of a lighted taper or candle held near flue pipe joints and the like will be drawn in to confirm the leakage. Leakage through the brickwork of a chimney is determined by a 'smoke test' (*see* Chapter 10).

Remedy All faulty flue pipe joints, and so on, should be made good, bearing in mind that fire cement is generally hard-setting and therefore liable to crack and fall out. There are non-hard-setting compounds available that are specially prepared for flue pipe joints, or alternatively soft heat-resisting rope, string or tape should be used according to the size and type of joint to be sealed. Non-hard-setting compound can then be applied if desired as a 'top dressing'. Leakage through brickwork can be remedied by an appropriate relining method as described in Chapter 8. If leakage appears around a register plate (notoriously difficult to seal), replacing this plate with a cast-in-situ or precast concrete lintel will allow proper seals to be formed.

Symptom B: No Updraught at All
B.1 Complete Blockage
If a flue is not kept properly cleaned, soot deposits in time can completely stop the draught; the same thing can happen when pieces of chimney pot, slate or brick fall down the flue. If deposits are allowed to build up in a flue and they become damp, they will get heavy and may fall from the flue wall and settle at a bend, causing a complete blockage.

Test The sweep's brush will reveal a blockage.

Remedy Try the sweep's rods first: if these are unsuccessful, a core ball lowered from the top may remove the obstruction, but in any case will indicate the exact position for opening the flue (*see* Chapter 10). Regular cleaning of the flue is important for safety, but will also make sweeping easier if there is less to be removed each time.

B.2 Cold Flue
An exposed chimney cools the flue gases rapidly and the draught is never as good as it is in a well-protected flue. Moreover, if the fire has been out for some time, or a flue damper or thermostat has been closed for several hours, the chimney draught may well be reduced sufficiently to cause smokiness.

Uninsulated cast-iron or other single-wall flue pipes should never be used outside a house as a chimney, as this is not allowed under the Building Regulations. The severe chilling of the flue gases, particularly when an appliance is burning slowly, often results in unsightly condensation. More seriously, it reduces the flue draught so much that dangerous fumes may be emitted from the appliance, because there is little or no flue draught to carry them safely up the flue.

Test   If gas or smokeless fuel is used, insert a lighted piece of newspaper in the flue; after a brief delay, the smoke should be carried up the flue. If bituminous coal or wood is burnt, this could result in a chimney fire and so the use of a burner on the grate is advised.

Remedy   Give the draught an initial 'boost' by lighting paper at the base of the flue so as to warm the air in the flue. A gas poker may be used for the purpose, but it is important to ensure that all jets are alight and that no unburned gas is escaping up the flue. Consideration should be given to insulating any long length of single-skin flue pipe or replacing these with an insulated chimney system.

In the case of a thermostatically controlled appliance, an escape of fumes is most noticeable in the morning after the thermostat has been shut for several hours overnight. Giving the boiler some work to do overnight will slightly increase the flue gas temperature and avoid fumes or spillage in the morning.

Chimney Terminal is in a High-Pressure Region

C.1 Chimney Top is in a High-Pressure Region
Figure 146 shows a chimney lower than a nearby object, in this case the roof, and on the windward side of that object. This may be a region of high wind-pressure. There is low-pressure (suction) on the lee side of the house. This results in air tending to be drawn out of the house on the suction side and being blown in on the pressure side. This can affect the chimney and produce smoke emission by either preventing the updraught, or, if the pressure is high enough, causing a flue reversal.

As this condition usually occurs only with the wind in a certain quarter, the situation of the chimney in relation to the roof will indicate the cause of the trouble. Smokiness normally occurs only when there are openings (doors or windows) on the low-pressure side but not on the high-pressure side.

Test   Open a door or window on the exposed pressure side of the property; this should equalize the pressures and restore updraught.

Remedy   This problem cannot be cured by any cowl as the pressure region will still exist. If possible, the chimney should be extended beyond the region of high pressure; this can be done experimentally in the first place by trying various lengths of sheet metal pipes to determine the extra height needed. In most cases, if it is 0.6–1m higher than the ridge or other object responsible for the pressure zone, no trouble will occur.

Unlagged pipes should not be used as permanent extensions, however, as they will chill the flue gases and possibly cause further trouble. The Building Regulations impose a definite limit on unsupported chimney height, and if the permitted extension does not prove sufficient, probably the only alternative is to prevent air being drawn out of the room towards the low-pressure region. This can only be done as a rule by keeping the door closed and seeing that it is reasonably airtight. A suction-inducing cowl will not normally help unless the top of the cowl is high enough to be out of the pressure area.

Downdraught and Difficult Site Conditions

D.1 Downdraught Due to Room Inlets (Doors, Windows, Ventilators) being in a Low-Pressure Region
This trouble occurs mainly with short chimneys (for example, bungalows and the top floor of a block of flats) and with inset open fires. 'Suction' in the room, although normally associated with the condition described in C.1, can sometimes cause smokiness regardless of chimney position. The illustration here shows a condition that in certain circumstances can be difficult to cure. The flow of air around a house creates regions of high and low pressure, which may

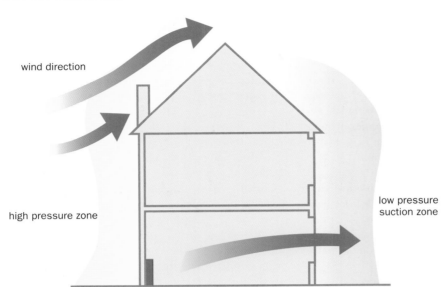

wind direction

high pressure zone

low pressure suction zone

Fig 146   A pressure zone on a chimney, caused by wind and a pitched roof.

cause downdraught if the position of doors and windows is such that suction exceeds pressure sufficiently to overcome the natural 'pull' of the flue. Never put air vents in areas such as alleyways between houses or passages, as the air speed through these is often greater and increases the risk of a suction zone forming.

Test   Produce sufficient aromatic test smoke in the room to show the movement of air; usually the smoke will drift towards doors and windows in low-pressure regions, thereby indicating that air is being drawn out of the room.

Remedy   As with most other draught troubles, some experimentation is usually necessary. The normal updraught can be increased by:

- replacing the existing fire with a closed room heater that reduces air flow considerably and maintains a higher temperature in the flue gases;
- fitting a draught-inducing cowl to increase up-draught;
- fitting a throat restrictor to reduce air flow and increase the temperature in the flue;
- draught-proofing doors and windows and so on that are adjacent to the suction areas.

D.2 Downdraught Due to Wind Currents
Downdraught can be due to the wind striking downwards onto the chimney top. Downward-striking wind currents often occur in the vicinity of chimney tops where there are higher objects nearby. Chimneys on the lee slope of a hill or in a valley may also suffer from this form of downdraught.

Test   Observe the position of the chimney top in relation to other, higher objects such as trees and buildings, or in respect of surrounding land contours. Note whether smoke has difficulty in issuing from the chimney pot and or is blown downwards.

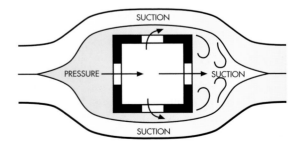

Fig 147   Suction and pressure caused by the wind on windows and vents.

Fig 148
Downdraught due
to wind currents at
the terminal,
caused by a higher
object or building.

downdraught

wind direction

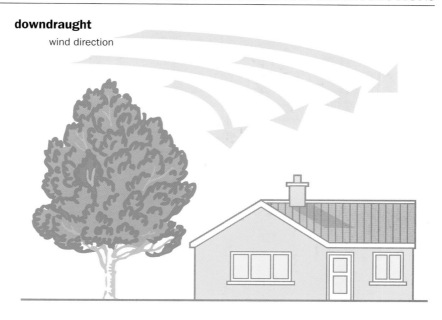

Remedy If it is impossible to raise the chimney terminal out of the turbulent zone, the simplest remedy is to protect the top of the chimney with an anti-downdraught or dovecote top.

Sometimes a cluster of chimney pots close together on the same stack may result in smoke and fumes from one flue passing down an adjacent one, particularly where fireplaces are near each other in communicating rooms. Raising the level of termination of one flue by fitting a tall pot is often sufficient to break the syphonage effect. If, however, adjacent flues (often in old houses) are not in use the problem is quickly resolved by capping the unused ones. Do not, however, do this in such a way that the flue cannot be easily opened up if it is likely to be needed later.

TOO MUCH DRAUGHT

Adequate chimney draught is necessary to induce sufficient air for combustion through the fuel bed and to clear the products of combustion from the appliance up the flue and disperse them to the atmosphere.

The amount of draught necessary for this varies with the type of appliance; for independent boilers the makers generally state the desired level of draught at the outlet of the appliance – usually in the region of 1.5–2.0mm (0.06–0.08in) water gauge. Since most well-constructed flues to which a closed appliance is connected produce a higher draught than this, the rate of combustion in a boiler may be higher than is desirable when the air control is open. If this is the case, it will often be too strong for the normal air control to check during periods when the boiler is idling, especially in windy conditions. For such circumstances an adjustable draught stabilizer should be fitted reasonably close to the outlet of the boiler and set by use of a draught gauge to give the optimum draught for the particular appliance. This means that in use the boiler will always operate under a steady 'pull' or draught – and more economically so than under varying and often excessive chimney draught. Due to the specialist equipment and knowledge needed this should only be undertaken by a heating or chimney engineer.

CONDENSATION

Water vapour is one of the products of combustion when any kind of fuel is burnt. Fuels with a relatively high hydrogen content produce more water vapour than others. As long as the water remains in its vapour state until it leaves the flue, no problems arise, but when the flue gases are cooled below a certain level, moisture will condense on the surfaces of the flue. Unfortunately, condensation often combines with sulphur or other compounds in the flue gases and with sulphate in brickwork to form weak acids, which, over a period of time, attack brickwork and mortar joints, causing damage to the chimney fabric. In some cases, condensation will even penetrate brickwork, staining plaster and interior decorations, particularly in rooms on upper floors. Other contributory causes are flues too large for appliances,

burning wet refuse such as damping down with tea leaves, and using very wet fuel stored in the open.

The remedy is to line the flue with an appropriately sized acid-resistant flue liner that is then insulated around with backfill. This prevents any further condensation from affecting the brickwork, and by reducing the heat loss from the flue the incidence of condensation is lowered. This condition is only likely to arise in houses built before 1966. In later houses all flues should be adequately lined with impervious material, but if the problem does occur in these flues it is an indication of more serious structural or liner problems.

Useful Addresses

Clear Skies Initiative
BRE Ltd
Building 17, Garstone
Watford
WD25 9XX
Tel 08702 430930

CORGI
Council for Registered Gas
   Installers
1 Elmwood
Chineham Business Park
Crockford Lane
Basingstoke
Hampshire
RG24 8WG
Tel 01256 372300

Guild of Master Sweeps
The Bungalow
London Road
Attleborough
Norfolk
NR17 2DE
Tel 01953 451322

HETAS Ltd
Heating Equipment Testing and
   Approval Scheme
PO Box 37
Bishops Cleeve
Gloucestershire
GL52 9TB
Tel 01242 673257

Institute of Plumbing
64 Station Lane
Hornchurch
Essex
RM12 6NB
Tel 01708 472791

Listed Property Owners' Club
Freepost
Hartlip
Sittingbourne
Kent
ME9 7TE
Tel 01795 844939

NACE
National Association of Chimney
   Engineers
PO Box 849
Metheringham
Lincoln
LN4 3WU
Tel 01526 322555

NACS
National Association of Chimney
   Sweeps
Unit 15, Emerald Way
Stone Business Park
Stone
Staffordshire
ST15 0SR
Tel 01785 811732

National Fireplace Association
6th Floor, McLaren Building
35 Dale End
Birmingham
B4 7LN
Tel 0121 200 1310

OFTEC
Oil Firing Technical Association
Foxwood House
Dobbs Lane
Kesgrave
Ipswich
IP5 2OQ
Tel 0845 6585080

SFA
Solid Fuel Association
7 Swanwick Court
Swanwick
Derbyshire
DE55 7AS
Tel 0845 601 4406

Further Reading

Approved Document J of the Building Regulations England and Wales

British Standards: BS 1251: 1987 – *Specification for Open Fireplace Components*

BS 6461 – *Installation of Chimneys and Flues for Domestic Appliances Burning Solid Fuel*

BS 8303 – *Installation of Domestic Heating and Cooking Appliances Burning Solid Mineral Fuels*

BS EN 12391 Chimneys – *Execution standards for metal chimneys*

Building Standards Scotland, Section F

HETAS: *Official Guide to Products and Services*

Solid Fuel Association: *Designing Solid Fuel into Homes*

Curing Chimney Troubles

170

Glossary

Air starvation   This is where a fire or flue is unable to get sufficient oxygen or air into the room to enable it to operate correctly.

Approved Document J   The document that provides the clauses that should be followed when installing a combustion appliance to be deemed to have satisfied the Building Regulations Section J.

Balanced flue   This can be horizontal or vertical and is an appliance where the air inlet and the flue outlet are both at the same outside location.

Bedding   The seating of a chimney pot, liner or other component onto a bed of mortar or jointing compound.

Bend   A single change in direction, usually from or back to the vertical.

Boiler   An appliance that has a heat exchanger where the heat is transferred to water for distribution around the property.

Canopy or hood   A metal gather used over a basket grate to collect the gases and draw them into the flue. It is usually visible and decorative.

Capital cost   The initial cost of an installation including the appliance, any flue and fuel storage needed.

Carbon monoxide (CO)   An incomplete molecule of gas formed when carbon is burned in an atmosphere that is short of oxygen. In the case of solid fuel, carbon monoxide will always be formed as it is not possible to mix the carbon and oxygen completely as it is with burning gas.

Cast in situ   This is where a liner or lintel is actually cast in the flue rather than cast somewhere else and then built into the flue later.

Central heating   The heating of the whole house from one central boiler or warm air heater.

Chimney   The structure that contains one or more flues.

Chimney stack   The part of the chimney structure that projects above the roof.

Chimney terminal   The top or capping on the chimney.

Combination boiler   A unit that combines two heat exchangers in the system, one to heat the water to be distributed around the radiators and one to heat the water that is then drawn directly from the taps.

Combustible material   Material that can catch fire when exposed to flames or heat.

Condensates   The liquids formed in a flue or on a heat exchanger when the temperature drops low enough for the moisture in the products of combustion to change from vapour to liquid.

Condensing boiler   A boiler that extracts so much heat from the products of combustion that they condense on the heat exchanger, giving out extra heat and working at higher efficiencies when this occurs.

Constructional hearth   The hearth required for a majority of solid fuel appliances. It is normally set in the floor and in the base of the fireplace recess to protect the structure from the high heat levels below these types of appliances.

Corbelling   Stepped brickwork that forms a gather or offset, sometimes used at the top of a chimney stack to form an oversail or drip course.

CORGI   Council for the Registration of Gas Installers; this is the competent persons scheme for the installation of appliances fired by gas, as well as the pipework and so on to transport the gas once it is within the property.

Decorative fuel effect   An appliance that imitates a real solid fuel or wood-burning appliance.

Decorative hearth   The decorative finish on the constructional hearth that creates a change in level from the floor to prevent combustible material such as carpets and rugs being placed too close to the appliance.

Dog basket/dog grate   In this case these are the same thing in that they are a basket in which the fuel is placed to burn, or a gas burner is placed to provide an artificial fire effect. They have extended front legs that come in front of and/or either side of the basket itself.

European Normative Standards   In the UK these will be identified as BS EN and be followed by a number. These are the standards that have been ratified for use around the European Community.

Factory-made chimney systems   As for prefabricated chimney systems, but normally used to describe those made from steel components.

Fireplace   Decorative surround that is placed around the appliance or fireplace opening.

Fireplace opening   In the case of an open fire this is the opening in the fire surround in which the fire sits.

Fireplace recess   Structure in which the appliance or fire will stand. It is often lined with a decorative finish or a British Standard fireback to form the position for the fire.

Flashing   The lead or similar material used to waterproof the joint between the chimney and any roof it passes through.

Flaunching (chimney pot)   The cement used to hold the chimney pot in place and form a shape that will shed rain water and condensation from the top of the chimney.

Flaunching (fireplace)   The cement shaping over the fireback or offtake pipe to create a smooth path for the flue gases.

Flexible stainless steel liners   These liners are sold in lengths that will go from the top to the bottom of the chimney with no joints. Their construction allows them to flex so that they can pass through bends in the flue. These can only be used to reline an existing flue as they are not acceptable as the primary liner in a new chimney.

Flue   The space formed by the chimney that allows the passage of flue gases.

Flue liner   The material used in the chimney stack which lines the chimney to form the flue.

Flue pipe   The pipe that connects the appliance outlet to the flue.

Flue spigot   The socket around the appliance outlet for the flue pipe to fit into.

Flue terminal   The terminal of the flue. It may be a chimney pot or a stainless steel terminal for a gas appliance flue.

Free-standing   An appliance that is not built into the chimney breast and is connected to the flue by a flue pipe. This can be stood inside a fireplace recess or not.

Gather   The space above the fireplace opening, and throat if one is fitted, which collects the gases into the base of the flue and gradually speeds them up. The more gradual and smoother the gather, the less resistance it causes to flue gas movement.

Heat exchanger   The part of the unit that collects the heat from the products of combustion and then provides them to the air or water that is to distribute them around the room or property.

HETAS   Heating Equipment Testing and Approval Scheme; the test and approvals body for solid fuel products and services, including appliances and competent people.

Inglenook fireplace   This is more than a large fireplace – it is an extension to the room with a smaller fireplace set in the back wall or a stove standing in it.

Inset   This is an appliance that is built into the chimney breast.

Midfeathers   The walls within a chimney that separate the flues within a multi-flue chimney (also known as withe or bridge walls in some parts of the UK).

Natural draught flue   A flue in which the draw of the chimney occurs due to natural phenomenon rather than a mechanical extract or pressurization system.

Offset   Part of the flue that takes the flue direction in a route that is not vertical. It consists of two bends and a straight section between them.

Offtake pipe A pipe that is used to direct the flue gases from the appliance to the flue in an inset installation.

OFTEC Oil Firing Technical Association; the competent persons scheme for the installation of appliances fired by oil, as well as the pipework and so on to store and transport the oil up to and within the property.

Oversail/drip course One or more courses of bricks that project out from the rest of the chimney stack to make rain water and condensation drip off the stack onto the roof to be washed away.

Part per million In the case of gases this is the number of molecules of the gas per million molecules of the air or products of combustion.

Period fireplace A description often given to a large fireplace.

Perlite A naturally occurring fireproof material that is treated in such a way that it becomes light and insulating.

Pointing The exposed mortar in the joints between bricks in a wall or chimney stack; at this point the mortar is shaped to enable the rain to run off without penetrating the joint. There are different pointing shapes depending on the location of the brickwork and its exposure to the wind and weather.

Prefabricated chimney system This is a chimney constructed from a series of factory-made components consisting of either steel tubes or concrete blocks.

Products of combustion When a fire or flame burns smoke and fumes are caused. These will contain a large number of different gases and solids and to encompass all these in one term they are called products of combustion.

Proposed European Standards In the UK these will be identified as prEN and be followed by a number. These are the standards that have still to be ratified for use around the European Community.

Pumice A volcanic rock used in some concrete chimney liners and block chimney systems.

Range cooker An open fire that has a range of ovens and flues within it to enable it to be used for cooking.

Rear outlet An appliance where the flue gases leave the appliance at the back.

Roof lantern A cover used over a hole in the roof when central hearths were present.

Room heater Often known as a stove, this is an appliance that has a glass panel in the front door; normally solid fuel/wood-burning, but now available in all fuels, even electric.

Sectional liner A flue liner that comes in sections and is placed into the flue, normally during construction, and which is jointed with acid-resistant jointing compound.

Smokeless fuel A solid fuel that when burnt will produce less than a regulated amount of soot and other compounds in the products of combustion. These will have some smoke and fumes but will be less visible than with coal. It will still contain dangerous compounds such as carbon monoxide and sulphur dioxide.

Solid mineral fuels These are the solid fuels extracted from the ground or made from these extracts. Most commonly known as coal or smokeless fuel.

Stove An appliance that has a glass panel in the front door, normally solid fuel/wood burning but now available in all fuels, even electric; also known as a room heater.

Superimposed or decorative hearth A raised section of floor intended to stop carpet and floor coverings getting too close to the appliance.

Throat An area of restriction in the flue just above the top of the fireplace opening. This restriction is to reduce the amount of air drawn out of the room up the chimney, while also speeding up the air movement to assist in the drawing of the smoke into the flue.

Top outlet An appliance where the flue gases leave the appliance at the top.

Vermiculite A naturally occurring fireproof material that is treated in such a way that it becomes light and insulating.

Void A large space within a flue or above a fireplace.

List of Contributors

I would like to thank all the companies and individuals below for their help in providing illustrations and information:

A & M Energy Fires
Pool House
Huntley
Gloucestershire
GL19 3DZ

Figure 107

A J Wells and Sons
Bishops Way
Newport
Isle of White
PO30 5WS

Figures 32 and 34

Ouzeldale Foundry Co. Ltd
Long Ing
Barnolswick
Colne
Lancashire
BB18 6BN

Figure 36

Brian Frost Chimneys
The Studio
Oxenbourne Farm
East Meon
Petersfield
Hampshire
GU32 1QL

Figure 96

Enginuity Developments
Lower Farm
Hayward Bottom
Hungerford
Berkshire
RG17 0PX

Figures 7 and 54

Chimneycrete Ltd
Victoria Lodge
Forest Moor
Knaresborough
North Yorkshire
HG5 8JY

Figure 97

Euroheat Distributors (HBS) Ltd
Unit 2
Court Farm Business Park
Bishops Frome
Worcestershire
WR6 5AY

Figures 21, 30 and 35

Hepworth Terracotta
Hazelhead
Stockbridge
Sheffield
S30 5HG

Figures 60, 63, 71, 72 and 75

HETAS Ltd
PO Box 37
Bishops Cleeve
Gloucestershire
GL52 9TB

Figure 48

Office of the Deputy Prime Minister
Eland House
Bresenden Place
London
SW1E 5DU

Figure 40

Parkray Ltd
The Britannia Suite
St James Buildings
79 Oxford Street
Manchester

Figure 40

Redbank Manufacturing Company Ltd
Atherstone Road
Measham
Swadlincote
Derbyshire
DE12 7EL

Figures 59 and 83

Rite-Vent Ltd
Crowther Estate
Washington
Tyne & Wear
NE38 0AB

Figures 81 and 82

Robys Ltd
The Old School House
Green Lane
Belper
Derbyshire
DE56 1BY

Figure 20, and cover illustrations

Stovax Ltd
Falcon Road
Sowton Industrial Estate
Exeter
Devon
EX2 7LF

Figures 8, 9, 26–29, 38, 39, 53 and 105

Scheidel Isokern
14 Haviland Road
Ferndown Industrial Estate
Wimborne
Dorset
BH21 7RF

Figures 74–78

The Chimney Company (SW) Ltd
14 Bowden Road
Ipplepen
Newton Abbot
Devon
TQ12 5QT

Solid Fuel Association
7 Swanwick Court
Alfreton
Derbyshire
DE55 7AS

Figures 14, 17–19, 44, 50, 51, 57, 65, 66, 103, 104, 106, 108, 109 and 123–148, and extracts form the 'Curing Chimney Troubles' leaflet

Villager Stoves
Lyme Regis Engineering Co. Ltd
Millway Industrial Estate
Axminster
Devon
EX13 5HU

Figure 33

Wöhler UK
13/14 EPOS House
263 Heage Road
Ripley
Derbyshire
DE5 3GH

Figures 124 and 125

www.dataplates.com
Victoria Lodge
Forest Moor Road
Knaresborough
North Yorkshire
HG5 8JY

Figure 48

Index